Climatology versus Pseudoscience

Climatology versus Pseudoscience

Exposing the Failed Predictions of Global Warming Skeptics

DANA NUCCITELLI

AN IMPRINT OF ABC-CLIO, LLC
Santa Barbara, California • Denver, Colorado • Oxford, England

Library of Congress Cataloging-in-Publication Data

Nuccitelli, Dana.
Climatology versus pseudoscience : exposing the failed predictions of global warming skeptics / Dana Nuccitelli.
pages cm
Includes bibliographical references and index.
ISBN 978-1-4408-3201-7 (alk. paper) — ISBN 978-1-4408-3202-4 1. Global warming. 2. Climatic changes. I. Title.
QC981.8.G56N93 2015
577.27'6—dc23 2014046795

ISBN: 978-1-4408-3201-7
EISBN: 978-1-4408-3202-4

19 18 17 16 2 3 4 5

This book is also available on the World Wide Web as an eBook.
Visit www.abc-clio.com for details.

Praeger
An Imprint of ABC-CLIO, LLC

ABC-CLIO, LLC
130 Cremona Drive, P.O. Box 1911
Santa Barbara, California 93116-1911

This book is printed on acid-free paper ∞

Manufactured in the United States of America

Contents

Figures

Preface

Like many people, *An Inconvenient Truth* first piqued my interest in climate change. Here was a subject about which I had heard very little, being presented as a major threat to humanity, with seemingly strong supporting evidence. As a scientist and a skeptic, I needed to learn more.

My background is in physics, with a bachelor's degree in astrophysics from the University of California at Berkeley, and a master's degree in physics from the University of California at Davis. I'm an environmental scientist at a private consulting firm, though contrary to the potential connotations of my job title, my work is in no way related to climate change. I investigate and clean up soil contamination, mostly from former military sites, so I'm not on the fictional "global warming gravy train." My initial interest in the subject was purely academic, but my scientific background gave me the tools to understand the basics of climate science. Although it's a complicated field of science, its fundamental concepts are not too difficult to understand, and I have tried to explain some of them in this book.

Intrigued by but skeptical of the arguments laid out in *An Inconvenient Truth,* I set out to learn about the subject for myself. Much of my free time over the past decade has been spent reading every climate-related text I can get my hands on, from peer-reviewed studies to books to blogs and mainstream media articles. I've found that the more I learn, the more concerned I become about the threats posed by rapid human-caused global warming.

To put it bluntly, climate change is probably one of the greatest threats the human race has ever faced. That's no exaggeration—we have tremendous infrastructure investments in cities and agricultural fields that depend upon the climate being relatively stable, and we are in the process of destabilizing it. Within the next few decades, the Earth's average temperature will be higher than it has been during the history of human civilization, over the past 10,000 years. We are conducting a grand and extremely risky global experiment, and all indications are that the results will be very, very bad if we continue on our current path.

Despite facing this immense threat, most Americans either are entirely unconcerned about climate change, in denial that humans are causing the planet to warm rapidly, or consider it very low on the list of priorities they need to address. This is understandable to some degree, since global warming is a long-term threat whose effects we will experience relatively slowly over a period of decades, whereas we also face many short-term problems like wars and economic recessions whose effects we experience immediately. People naturally prioritize the immediate threats first, but unfortunately they pale in comparison to the threats posed by climate change.

In my research, I've found that the American politicization of science and journalism has become a major roadblock in the way of widespread public understanding and acceptance of the threat posed by global warming. Most political liberals understand that the planet is warming and humans are the primary driving force, but most American political conservatives don't believe either of these well-established facts. The scientific evidence and data of course have no political affiliation, but ideological biases can make people reject ideas that they don't want to believe.

Because global warming is such a huge problem, and because essentially every part of our economy is so reliant on the fossil fuels that are causing the problem, the solution must necessarily involve government action. However, political conservatives oppose "big government," and thus many also oppose the main solutions to the climate change problem. The easiest way to justify opposition to a problem's solutions is to deny that the problem exists, as we saw when the tobacco industry denied the health impacts of smoking. The tobacco industry hired its own scientists to argue that smoking was safe and to sow doubt in the public's minds, just as fossil fuel industries have done with human-caused global warming.

The good news is that a bipartisan solution has been proposed—a small government, free-market, revenue-neutral carbon tax. In this system a fee is applied to carbon emissions at the source (at the well, mine, or port). This causes energy and gas prices to rise, but 100 percent of the revenue generated is returned to taxpayers. None is used to increase the size of government, and because the revenue is returned to the taxpayers, it also has minimal financial impact on people. In fact, about two-thirds of people will receive a regular rebate check that's larger than their added energy costs. Only the biggest carbon polluters will pay more than they receive in return. The legislative proposal starts with a low carbon price that rises every year, so people can adjust their purchasing decisions knowing that they can save more money by buying products that will lower their carbon emissions. It's a simple proposal to give people a financial incentive to reduce their carbon pollution without significantly hurting their wallets or the economy. Many conservatives thus support the proposal, but so far no Republican politicians currently in office have been willing to support it. Unfortunately, the political atmosphere remains too partisan, and many conservatives view climate science as a liberal issue that they must reject.

While researching the many climate myths that bounce around the Internet and media, I came upon the website Skeptical Science, run by Australian John Cook. Like me, Cook had a background in physics and a curiosity and skepticism about climate science. Cook started to examine some of the claims made by so-called climate change "skeptics" (who in general are not really skeptical, but rather reject mainstream climate science due to their ideological biases; hence, I prefer the term "contrarians") and how they stacked up against the peer-reviewed scientific literature. Cook found that in almost every case, the contrarian arguments were fundamentally flawed and contradicted by the body of scientific research, and he began to assemble a database of these myths and what the peer-reviewed science said about them, which became SkepticalScience.com.

Because of the quality and usefulness of the site, and the sheer number of myths that the climate change contrarians have come up with, Cook soon had a large readership requesting more mythbusting than he could keep up with. He invited other climate geeks to help him build his database, and I began contributing to Skeptical Science in September 2010. I've become a regular contributor, having written hundreds of articles and debunked dozens of the over 200 climate myths in the Skeptical Science database.

In December 2012, the British newspaper *The Guardian* announced that it would be creating a network of international environment blogs. *The Guardian* indicated that it would set up the blogs and then "turn over the keys" to the individual environment bloggers. *The Guardian* had already established itself as one of the world's best newspapers, especially in terms of environment and climate reporting, so I jumped at the opportunity to apply for this new blog network. So did 800 other people.

Ultimately, *The Guardian* selected a dozen or 1.5 percent of the applicants for its environmental blog network. Despite being a British newspaper, *The Guardian* had an extensive global readership and wanted its new environment blogs to reflect that international flavor. Hence, it chose applicants from Australia, the United States, Africa, China, Canada, India, and South America. Fortunately, *The Guardian* had previously republished several of my blog posts from Skeptical Science, so it was already familiar with my work. When I heard that it was also considering an application from John Abraham, a scientist at the University of St. Thomas in Minnesota with whom I had previously collaborated, I spoke with John and we offered to coauthor a blog. In May 2013, *The Guardian* established its new environmental blog network, including a blog run by myself and John Abraham. We named our blog "Climate Consensus—the 97%" for reasons that will become clear in this book.

Several factors motivated me to write this book. Science historian Naomi Oreskes recently criticized the climate science community for being reluctant to take credit for its many successful predictions.[1] I've also found that climate change contrarians are held to a much different standard than mainstream climate scientists. If a climate scientist makes a minor error, it's often blown out of proportion to create a manufactured scandal. On the other hand, climate contrarians are allowed to constantly make false and misleading claims and are rarely held accountable for their constant errors. Numerous contrarians have told us for years that global warming is nothing to worry about and the planet will start to cool any day now. The planet continues to warm, and yet people continue to listen to these same contrarians. It seems as though their credibility is indestructible. They can be wrong about every single one of their climate-related assertions for decades on end, and yet they continue to be treated as credible "skeptic" experts by the mainstream media, who seem desperate to find those few seemingly credible scientists to "balance" the vast majority of climate experts who agree that humans are the main cause of global warming.

Unfortunately, we're rapidly running out of time to realistically be able to achieve sufficient greenhouse gas emissions cuts to avoid very dangerous climate change. Doing so will require international cooperation, and the United States, as one of the wealthiest nations and the world's largest historical emitter, is a key to international negotiations. In international climate talks, both China and India have agreed to commit to emissions cuts, but first they need the big historical emitters like the United States to start taking serious steps to address it. It only makes sense for the countries that have contributed the most to the problem and become wealthy by exploiting fossil fuels to take the first steps to resolve it.

However, the odds of the United States taking serious action to reduce our greenhouse gas emissions are not promising because the climate disinformation campaign has been so effective in misinforming the public. A large percentage of Americans think there is major disagreement between climate scientists on the subject and that "the science isn't settled." In this book, we'll see that scientists have understood the basic science behind global warming for over a century, climate scientists have been accurately predicting global warming for over 40 years, and their predictions have been getting better and better as we grow to understand more nuances about how the global climate operates. Those predictions show that we're rapidly running out of time to address the problem. The good news is that we still have time, and we have all the technology we need to solve the problem. All that's missing is the will.

The stakes are exceptionally high. The climate is already changing at a dangerously rapid rate, and the faster it changes, the more difficult it will be for species to adapt to those changes. As we'll see later in the book, due to a number of human influences including climate change, species are already becoming extinct at a rate that suggests we could be headed toward a mass extinction event—only the sixth in the Earth's entire history.

The first chapters of this book discuss some important discoveries in the history of climate science, to lay out the fundamental physics on which the science is based, which has been established for a very long time. The main purpose of the book is to test the various global warming and cooling predictions made by mainstream climate scientists and climate change contrarians over the years, to see who has been right, and, as Oreskes suggested, to give those who have made accurate predictions their due credit. Some of the results may surprise you, as they did me.

The final chapter of the book examines what the most accurate predictions to date show that we have in store for the future and looks at some possible options to alter our current course. Ultimately, the purpose of this book is to examine global warming and cooling predictions with a truly skeptical eye and see which stand up to scrutiny.

Acknowledgments

Thanks to John Cook, of SkepticalScience.com for creating a great website through which I was able to learn and write about climate science, who encouraged me to write this book and who blinged up its graphics. And to my wife Betsy, for putting up with me sitting at my computer writing it for so long.

1

The Origins of Climate Science and Denial

People often mistakenly think of climatology (the study of how the Earth's climate functions) as a young field of science. Although climate scientists have only been creating complex models to simulate the Earth's climate in great detail for a few decades, we've understood the basic mechanisms that drive changes in the Earth's temperature and climate for well over a century.

In this chapter, we'll examine some of the early scientific discoveries that helped scientists understand the Earth's greenhouse effect and its importance in driving changes in the global temperature and climate throughout the Earth's history, but especially since humans began pumping billions of tons of carbon pollution into the atmosphere every year.

These groundbreaking scientific discoveries became the foundation of the field of research we now know as climatology. The first of these key scientific breakthroughs came nearly two centuries ago.

1820s

Jean Baptiste Joseph Fourier was a French mathematician and physicist in the late 18th and early 19th centuries. His ideas established our first understanding of how human activity can impact the Earth's climate. In the 1820s, Fourier calculated that, given the size of the Earth and its distance from the sun, if the planet were warmed solely by solar energy, it should be much colder than it is.

Fourier suggested a few mechanisms that could account for the Earth's extra warmth. One of his ideas, based on an experiment by fellow physicist Horace-Bénédict de Saussure, was that air might absorb heat rising from the Earth's surface, preventing it from escaping into space, thus warming the planet.

Fourier's idea is widely recognized as the first proposal of what we call "the greenhouse effect."[1] This fundamental scientific basis behind the human-caused global warming theory was established nearly two centuries ago, and the physicists who followed Fourier continued to build upon his theory.

Consistent with Fourier's proposal, greenhouse gases are molecules that absorb radiation at infrared wavelengths. Radiation from the sun arrives on Earth mainly in the visible and ultraviolet wavelengths. The planet absorbs this radiation and then reradiates it away as heat in the infrared wavelengths. Greenhouse gases in the atmosphere then absorb this infrared radiation and reemit it in all directions. Some of it gets sent out into space, but some of that energy is directed back toward the Earth's surface. It's similar to adding another blanket to your bed; less heat is allowed to escape, and this keeps the layers below warmer.

Despite our nearly 200-year understanding of the greenhouse effect, in recent years, it has become very popular for climate contrarians to argue that carbon dioxide can't be responsible for global warming because there's so little of it in the atmosphere. Carbon dioxide makes up 400 parts per million (ppm) of the atmosphere, which is just 0.04 percent. Many climate contrarians have argued that, as such, a small trace gas as carbon dioxide certainly can't be dangerous.

There are a few reasons why this argument is fundamentally incorrect. First, the concentration of a substance isn't enough information to determine whether it poses a threat.[2] For example, arsenic in drinking water is considered a threat to human health at 0.01 ppm (0.000001 percent). Dioxins in soil are considered a health threat at about one part per trillion (0.0000000001 percent). In the United States, it's against the law to drive with a blood alcohol level of just 0.08 percent. So even trace amounts of certain substances can be dangerous if, for example, they're highly toxic. In the case of carbon dioxide, its effectiveness as a greenhouse gas (a.k.a. its "global warming potential") is another important piece of information. There are a number of greenhouse gases with higher global warming potentials than carbon dioxide, but carbon dioxide is a significantly more effective greenhouse gas than water vapor, which is the only greenhouse gas more common in the Earth's atmosphere than carbon dioxide.

The second major flaw in this argument is that the greenhouse effect cannot be diluted. Ninety-nine percent of the molecules in the Earth's atmosphere are not greenhouse gases, which is why the amount of carbon dioxide in the atmosphere seems so small. However, these non-greenhouse gases don't make the greenhouse effect any weaker. Doubling or halving the amount of nitrogen, oxygen, and argon in the atmosphere would have very little effect on the amount of infrared radiation absorbed and reemitted by its greenhouse gases.

Many climate contrarians emphasize the atmospheric percentage of carbon dioxide in order to misinform the public, because it *sounds* small.[3,4,5] However, what really matters is the total amount of greenhouse gases in the atmosphere; the percentage tells us very little, and focusing on that number can be very misleading.

As Jean Baptiste Joseph Fourier realized nearly 200 years ago, the Earth would be a much colder place if not for the greenhouse gases in the atmosphere.

1859

John Tyndall was a prominent British physicist in the mid-19th century and also an avid Alpine mountaineer. While climbing in the Alps, Tyndall studied glaciers. In the process, he became convinced that tens of thousands of years ago, all of northern Europe was covered by ice. For this theory to be true, Tyndall was forced to explain how the climate could warm dramatically enough to make so much ice disappear. Jean Baptiste Joseph Fourier's greenhouse effect provided a possible explanation.

In 1859, Tyndall set up laboratory experiments to measure the amount of heat absorbed by various greenhouse gases and was the first scientist to measure the greenhouse effect.[6] Tyndall correctly concluded that because of the large amount of water vapor in the atmosphere, it is responsible for the biggest fraction of the greenhouse effect on Earth. Tyndall's experiments also showed that carbon dioxide is an effective greenhouse gas and plays a significant role in the Earth's greenhouse effect.[7]

Tyndall also arrived at another important conclusion: that the planet would be much colder at night if not for the greenhouse effect. At night, when there's no solar radiation bombarding the Earth's surface, greenhouse gases continue absorbing heat, thus keeping the surface relatively warm.

Today, many of the lines of observational evidence supporting the human-caused global warming are not very widely known. Because of this, many people don't believe that a convincing case linking human greenhouse gas emissions to the current global warming exists. Although most people realize that the planet is warming, many believe it's just as likely that the warming could be natural rather than human in origin.

However, we can apply Tyndall's conclusion regarding greater warming at night than during the day to test whether greenhouse gases are responsible for global warming. If the current warming is due to an increased greenhouse effect, and the greenhouse effect plays a larger role at night, then we would expect to see more warming at night than during the day over the past century. However, if, for example, the warming were due to increased solar activity, we would expect to see greater warming during the day, when solar energy is bombarding the planet's surface.

Climate scientists have examined the global warming trends at night and during the day, and the observational data confirms that, consistent with man-made global warming, the planet has warmed more at night than during the day.[8] This is one of many key observational "fingerprints," which demonstrate that the current warming is being caused by human influences.[9]

Perhaps the single most convincing of these fingerprints is the cooling of the upper atmosphere. As Fourier suggested, greenhouse gases prevent some heat from escaping into space by absorbing and reemitting it in all directions. In addition, most greenhouse gases reside in the lower atmosphere (the troposphere). Thus, when we release more carbon dioxide into the atmosphere, more heat is trapped in the lower atmosphere. As a result, if global warming is caused by this increased greenhouse effect, we expect to see less heat reaching the upper layers of the atmosphere, causing them to cool. Once again we would expect to see the opposite if the sun were causing global warming, because the increased solar radiation would warm all layers of the atmosphere. In fact, aside from an increased greenhouse effect, there aren't very many ways to explain why the upper atmosphere would cool as the Earth's lower atmosphere and surface warms.

Satellites and weather balloon measurements have observed exactly that the upper atmosphere has cooled while the lower atmosphere has warmed.[10,11] These are just two of the many fingerprints of man-made global warming observed by climate scientists. Combined with fundamental physics,[12] this scientific evidence quite clearly demonstrates that the current global warming is predominantly caused by the increase in atmospheric greenhouse gases, consistent with John Tyndall's experiments over 150 years ago.

There are also many lines of evidence proving that the increase in atmospheric greenhouse gases is due almost entirely to human activities (mainly burning fossil fuels and deforestation).[13] The clearest line of evidence involves simple accounting. Humans are emitting about 30 billion tons of carbon dioxide per year, but the amount in the atmosphere is only increasing by about 15 billion tons per year. Half of our carbon emissions are absorbed by plants and the oceans, and the other half ends up in the atmosphere. Simply put, our carbon emissions have to go somewhere.

1896

Svante Arrhenius was a Swedish physicist and chemist in the late 19th and early 20th centuries, who won the Nobel Prize for chemistry in 1903. In the late 1890s, Arrhenius set out to quantify how much greenhouse gases warmed the planet. After spending a year performing thousands of calculations using readings taken by American astronomer Samuel Langley, who had tried to work out how much heat the Earth received from the full moon, Arrhenius concluded that doubling the amount of carbon dioxide in the atmosphere would cause the planet to warm 5 to 6 degrees Celsius (°C), or about 9 to 11 degrees Fahrenheit (°F).[14] Amazingly, this estimate is within a factor of two of the estimates by today's climate scientists, who believe doubling atmospheric carbon dioxide will cause 1.5 to 4.5°C (2.7 to 8.1°F) of global surface warming.

As a result of this work, some have referred to Arrhenius as "the father of climate change."[15] Arrhenius also concluded that if humans continued to burn coal at the rate it was being burned in 1896, carbon dioxide levels would steadily increase, rising by about 50 percent in 3,000 years. In a 1908 paper, Arrhenius concluded that this rise in carbon dioxide would have a number of benefits, including possibly preventing the next ice age.[16]

> By the influence of the increasing percentage of carbonic acid in the atmosphere, we may hope to enjoy ages with more equable and better climates, especially as regards the colder regions of the earth, ages when the earth will bring forth much more abundant crops than at present, for the benefit of rapidly propagating mankind.

Of course, our rate of fossil fuel consumption has increased considerably since Arrhenius's time. In fact, over the past 150 years, the amount of carbon dioxide in the atmosphere has increased 40 percent. This is a rate of increase nearly 20 times faster than Arrhenius anticipated, and if we continue on our current path, atmospheric carbon dioxide will double preindustrial levels by the mid-21st century.

Arrhenius's optimistic perspective on the carbon dioxide increase is mirrored by many of today's climate contrarians, who argue that it will result in many positive consequences, like improved plant growth and fewer cold-related deaths. Unfortunately, the world is not so simple. Over the past few decades, the total amount of vegetation has increased globally, as Arrhenius anticipated.[17] However, carbon dioxide is not the only factor that affects plant growth.

In a real greenhouse, if we keep temperature and moisture steady at ideal levels and increase the amount of available carbon dioxide, plants will grow larger. On the other hand, when we increase the amount of carbon dioxide in the Earth's atmosphere, we can't hold these other factors steady. The planet warms, resulting in more evaporation, leaving less moisture in the soil for plants. More moisture is transported to the atmosphere, which results in stronger storms.

Global warming also tends to occur fastest near the poles. This is due in large part to the amount of ice there; when it melts, instead of being covered in reflective ice, the dark oceans below are uncovered. Darker surfaces absorb more sunlight, amplifying the existing warming in the region. Because the high latitudes warm most quickly, this decreases the temperature difference in the atmosphere between the poles and the equator. This in turn makes storms move more slowly, and because of the increased evaporation, those storms hold more water.

As a result, areas that are already wet are generally expected to get wetter, because they'll be hit by stronger, slower-moving storms. Areas that are already dry are generally expected to get drier, because the storms will have dumped most of their precipitation before they reach the already dry areas. For this reason, global warming results in more "extreme weather" such as floods and droughts, neither of which is beneficial for plant growth. Recent studies have concluded that the severity of droughts will increase substantially in many regions around the world over the next century if we continue on our current high-carbon emissions path.[18]

There are many other factors that impact plant growth as well. The bottom line is that continuing to increase atmospheric carbon dioxide will benefit some plant species and harm others. We're effectively conducting a grand global biological experiment, and a dangerous one at that.[19]

There are a number of other adverse consequences to increasing atmospheric carbon dioxide as well.[20] Ocean acidification is one such negative consequence that is all too frequently overlooked. The oceans absorb much of the additional carbon that humans pump into the atmosphere, and as a result of a chemical reaction, this causes the pH of the oceans to decrease.[21] This ocean acidification is happening at a very fast rate. Endorsed by 70 academies of science from around the world, a June 2009 statement made by the InterAcademy Panel on International Issues is as follows:[22]

> The current rate of change is much more rapid than during any event over the last 65 million years. These changes in ocean chemistry are irreversible for many thousands of years, and the biological consequences could last much longer.

As the oceans become more acidic, it becomes more difficult for marine life like corals and shellfish to form the hard shells necessary for their survival. Coral reefs provide a home for more than 25 percent of all oceanic species, so the damage done to corals through ocean acidification can result in a domino effect. In addition, tiny creatures called "pteropods" located at the base of many oceanic food chains can be seriously impacted by ocean acidification. The degradation of these species at the foundation of marine ecosystems could lead to the collapse of these environments with devastating implications to millions of people in the human populations that rely on them.[23]

Ultimately, the main problem is that humans are causing the global climate to change so rapidly that it will be difficult for many species to adapt quickly enough to the changing environment. There will certainly be some positive consequences to come out of these changes, but scientific research indicates that in all likelihood, the negative consequences will far outweigh the positives.

1900

Another Swedish physicist Knut Ångström may reasonably be considered the first climate change skeptic (although his motivations were very different from those of today's contrarians; Ångström was a genuine skeptic). Ångström was unconvinced by Arrhenius's calculations regarding the warming effects of increased atmospheric carbon dioxide, and he set up a laboratory experiment to test his conclusions. Ångström asked his assistant to measure the passage of infrared radiation through a tube filled with carbon dioxide. The assistant put in a somewhat lower carbon dioxide concentration than that found in the Earth's atmosphere and then cut the amount by one-third. The amount of radiation passing through the tube changed only 0.4 percent between the two experiments, and thus Ångström concluded that carbon dioxide was such an effective greenhouse gas that the greenhouse effect became saturated even at low concentrations. Therefore, Ångström concluded that adding more carbon dioxide to the atmosphere would not increase the greenhouse effect or cause any further warming.[24]

Ångström's conclusion has more recently been referred to as "the saturated gassy argument."[25] The primary flaw in the argument is that the Earth's atmosphere behaves as a series of many different layers, whereas Ångström treated it as one large slab. Even if the greenhouse effect were saturated in the lower layers of the atmosphere, the amount of infrared radiation reaching space would still depend on the greenhouse effect in the higher layers of the atmosphere. Some of the infrared radiation absorbed by the greenhouse gases in the lower layers of the atmosphere is reemitted upward, where it's likewise absorbed and reemitted by the greenhouse gases in the higher layers. Adding more carbon dioxide to these higher layers will thus make it more difficult for the infrared radiation to escape into space, and the planet will warm as a result. Returning to our previous analogy, adding more carbon dioxide to the atmosphere is like putting more blankets on a bed.

Indeed, climate scientists have used satellites to measure a decrease in the amount of infrared radiation leaving Earth,[26] and instruments on the surface to measure an increase in the amount of infrared radiation reaching the Earth's surface.[27,28] These measurements prove that the greenhouse effect is not saturated.[29] Arrhenius was right, and Ångström was wrong.

Despite the fact that Arrhenius's findings are over a century old, and as the next chapters will show, subsequent research proved him right over a half century ago, many of today's climate change contrarians continue to make Ångström's "saturated gassy argument."

For example, Senator James Inhofe (R-OK) is the most vocal global warming contrarian in the entire U.S. government, having announced on the U.S. Senate floor in 2003 that global warming is "the greatest hoax ever perpetrated on the American people."[30] Marc Morano was Senator Inhofe's communications director from 2006 to 2009 and more recently has created a website called "Climate Depot" to spread disinformation about global warming and climate science in general.

In 2007, while acting as Senator Inhofe's communications director, Morano gave an interview in which he made the same faulty carbon dioxide saturation argument as Ångström had made over a century earlier.[31] In fact, as discussed in chapter 5 of this book, the saturated gassy argument even continues to make its way into scientific articles to this very day, though not in articles written by climate scientists.

The fact that this long-debunked argument is still made today goes to show that climate contrarianism is based on a failure to consider the full body of scientific evidence and research, and a failure to learn from the past. As philosopher George Santayana put it,

> Those who cannot remember the past are condemned to repeat it.[32]

2

The Growth and Development of Climate Science

Our scientific understanding of the workings of the Earth's climate grew slowly but surely through the early and mid-20th century. There was relatively little research in the field of climatology in the early 1900s, but that began to change toward the middle of the century as our technology improved and the military began developing weapons that were based on science related to the Earth's greenhouse effect.

Once this military research interest kicked in, some important climate research papers were published, and the scientific field began to develop and advance at a rapid pace.

1938

Although we now recognize the fundamental errors in Knut Ångström's arguments about greenhouse effect saturation, most scientists in the early 20th century found them convincing. Up until the mid-20th century, there were only a few scientists who continued to build upon Arrhenius's work, trying to determine how much global warming humans could cause by burning fossil fuels.

Guy Callendar was one of those scientists. Callendar was an English steam engineer and inventor who also studied the greenhouse effect. Callendar compiled various measurements of temperature and atmospheric carbon dioxide and concluded that both were rising. Although atmospheric carbon dioxide measurements at the time were a challenge because they could be biased by local carbon dioxide sources like factories, Callendar estimated that atmospheric carbon dioxide had increased by 10 percent over the prior 100 years.[1] This estimate has turned out to be remarkably accurate, as current measurements place the atmospheric carbon dioxide increase over this period at about 9 percent.[2]

In 1938, Callendar published a paper in which he argued that the global warming in the early 20th century was caused by the increased greenhouse effect from this rise in atmospheric carbon dioxide.[3] We now know that most of the global warming during this period was caused by natural effects, such as an increase in solar activity and low volcanic activity (volcanoes cause short-term cooling effects by releasing particulates into the atmosphere that block sunlight). However, human carbon dioxide emissions did also play a significant role in this warming, and Callendar's work paved the way for future improvements in our understanding of the human role in climate change.

Between 1910 and 1940, the Earth's average surface temperature warmed about 0.4°C (0.7°F). Atmospheric carbon dioxide increased only approximately 5 percent during this time frame, so although it played a role, causing approximately one-quarter to one-third of this early 20th-century global warming, the warming was mostly due to natural causes.[4]

Some climate contrarians argue that because the early 20th-century warming was mainly natural, that means the current global warming could also be mostly natural.[5] It's true that hypothetically current warming *could be* caused by natural effects, but in reality, it's not. This is a logical fallacy, like arguing that because people died of natural causes before cigarettes were invented, smoking doesn't kill people now.

In the early 20th century, there was a fairly significant increase in solar activity, and thus in the amount of solar energy reaching the Earth's surface. This increase in solar activity was enough to cause approximately 0.1°C (0.2°F) global surface warming, similar to the amount of warming caused by the increased greenhouse effect during that same period. As previously noted, this was also a period with few significant volcanic eruptions, and some of the warming was probably also caused by ocean cycles which can impact short-term temperatures over a period of up to a few decades.

However, since 1950, solar activity has been flat on average, even declining slightly. Volcanic activity has increased somewhat over this period. So the natural factors that caused most of the early 20th-century global warming have actually had a small cooling effect since the mid-20th century.[6] Climate scientists have examined all factors known to impact the Earth's temperature, and since 1950, almost all of the observed global warming is due to the increased greenhouse effect from humans burning fossil fuels.

1956

In the 1950s, as a consequence of World War II and the Cold War, there was a significant increase in U.S. government funding (particularly from the military) for studying the Earth's atmosphere and oceans. For example, the military funded research dealing with infrared radiation in the atmosphere to develop heat-seeking missiles, and absorption of infrared radiation in the atmosphere is what causes the greenhouse effect.

In 1956, Canadian-American physicist Gilbert Plass made use of early electronic computers to calculate the warming effect of the increased greenhouse effect caused by adding carbon dioxide to the atmosphere. Plass published a paper concluding that Ångström was wrong regarding the saturation of the greenhouse effect and that doubling the amount of carbon dioxide in the atmosphere would cause between 2.5 and 3.8°C (4.5 to 6.8°F) global warming, depending on how cloud cover responds to that warming.[7] Clouds tend to reflect sunlight and thus generally have a cooling effect, although they also cause warming via the greenhouse effect; on average, the cooling effect is larger.

In another paper published in 1956, Plass warned of the dangers associated with conducting a grand global climate experiment by dramatically increasing the amount of carbon dioxide in the atmosphere.[8]

> If at the end of this century, measurements show that the carbon dioxide content of the atmosphere has risen appreciably and at the same time the temperature has continued to rise throughout the world, it will be firmly established that carbon dioxide is an important factor in causing climatic change.

The total amount of global warming (or cooling) caused by an energy imbalance (a.k.a. "radiative forcing") on Earth is known as "climate sensitivity." For example, we know from fundamental physics that by itself, if we double the amount of carbon dioxide in the Earth's atmosphere, the corresponding global energy imbalance caused by the increase in the greenhouse effect will be enough to warm temperatures on the planet's surface by about 1.2°C (2.1°F).

However, when the planet warms, other effects that influence its temperature are also triggered. These are known as "feedbacks," and they can either amplify or dampen the warming or cooling effect. For example, when the atmosphere warms, it can hold more water vapor. As another greenhouse gas, increased amounts of atmospheric water vapor also increase the greenhouse effect, causing further warming (a positive, amplifying feedback). Another example is ice, which is very reflective. When the planet warms and ice melts, it reveals the darker surface below, making the planet less reflective (this reflectivity is also known as "albedo"), causing the Earth to absorb more solar radiation and warm further (another positive feedback).

Cloud cover is probably the trickiest feedback to predict, because not only is it difficult to determine how different types of clouds will respond to a warming world, but clouds have both warming (by trapping heat) and cooling (by reflecting sunlight) effects. High-elevation clouds tend to have a net warming effect by trapping more heat, whereas low-elevation clouds tend to have a net cooling effect by reflecting more sunlight.

Water vapor and melting ice are two of the largest feedbacks, and because they're both positive and amplifying, the increased greenhouse effect will cause more global warming than would result from carbon dioxide alone. Climate sensitivity is commonly measured as the total amount of global warming that will result from a doubling of atmospheric carbon dioxide and is a very important concept.

Climate scientists have found many different ways to estimate the Earth's climate sensitivity.[9] They have examined past climate changes over hundreds of thousands to millions of years, as determined from measurements of data obtained in ice cores and other geologic records (this is called "paleoclimate" research). They have examined more recent climate changes over the past few thousand years, and over the past 150 years, during which time we have the most accurate measurements. Climate scientists have also examined the climate's response to recent large volcanic eruptions (which cause short-term cooling by blocking sunlight) and changes in solar activity, which cause a large enough energy imbalance to allow for a measurement of the Earth's temperature response. They also run climate models that simulate the Earth's climate, to see how much warming they predict in response to a given energy imbalance.

Remarkably, all of these different methods of estimating climate sensitivity are in good agreement with each other, and with Plass's 1956 estimates. They all indicate that if the amount of carbon dioxide in the atmosphere doubles, once the Earth's climate reaches a new energy equilibrium and stops warming, its average surface temperature will have increased between 1.5 and 4.5°C (2.7 to 8.1°F).[10]

For a short period of time between about 2012 and 2014, a few studies were published that were exceptions to this rule. These studies combined recent observational data with simple climate models and seemed to conclude that the climate sensitivity was somewhat lower than estimates based on geologic records or complex climate models. The so-called energy balance studies began arriving at an estimate that the planet's surface would warm closer to 2°C (3.6°F) in response to a doubling of carbon dioxide in the atmosphere,[11,12,13] whereas the other methods consistently arrived at best estimates closer to 3°C (5.4°F). It's important to note that the results of these different approaches all overlapped within their margins of error, but their best estimates were nevertheless different.

For a while these "energy balance" results were a bit mysterious—why didn't they agree with the other methods to estimate the sensitivity of the Earth's climate? A paper published in March 2014 by Drew Shindell at the National Aeronautics and Space Administration (NASA) suggested an explanation.[14] Shindell had coauthored one of the aforementioned papers that arrived at a lower best estimate of the climate sensitivity, but subsequently believed he found a fundamental error in the approach that he and his colleagues took in that study.

In their approach, the scientists had assumed that the various factors causing global energy imbalances were all equally efficient at changing the Earth's surface temperature. Any effect causing an energy imbalance of 1 Watt over a square meter of the Earth's surface was assumed to cause the same amount of global warming. However, James Hansen, a world-renowned climate scientist also from NASA, had first noted in a 1997 paper that climate influences have a bigger effect on global temperatures in the Northern Hemisphere.[15]

> A forcing at high latitudes yields a larger response than a forcing at low latitudes. This is expected because of the sea ice feedback at high latitudes and the more stable lapse rate at high latitudes.

In his 2014 study, Shindell realized that while greenhouse gases are well mixed throughout the atmosphere, aerosols and ozone are more concentrated in the Northern Hemisphere. Using several different climate models, he found that the temperature response to ozone and aerosol changes was thus up to 50 percent larger than the response to changes in greenhouse gases. His results suggested that this discrepancy could potentially explain why the climate sensitivity best estimates in recent energy balance studies were arriving at lower values than other approaches. It seemed as though they were the odd ones out for a reason.

Texas A&M University scientists John Kummer and Andrew Dessler published a paper following up on Shindell's study in May 2014.[16] Dessler described their approach to me.

> I view my paper as a follow-on to Shindell's paper. What he showed in his paper was that climate models respond more strongly to forcing from aerosols and ozone. What we show our paper is that if we take his result, and re-analyze the 20th-century observational record then we get a higher climate sensitivity than [studies] which assumed that all forcing was equally effective. Taking efficacy into account, our climate sensitivity is right in the middle of the values derived from other sources. So this allows us to bridge the gap between the various estimates of climate sensitivity and converge on a value around 3°C.

In short, Shindell showed that according to models, the climate is significantly more sensitive to changes in aerosols and ozone than greenhouse gases, perhaps by as much as 50 percent. Kummer and Dessler showed that if the climate is 33 percent more sensitive to changes in aerosols and ozone, then the energy balance estimates are right in line with those derived from historical climate changes and global climate models, with a best estimate of 3°C (5.4°F) warming in response to a doubling of atmospheric carbon dioxide. If their results are correct, then there is once again near universal agreement using all available methods that global surface temperatures will warm between about 2 and 4.5°C (3.5 to 8.1°F) in response to the increased greenhouse effect from doubled atmospheric carbon dioxide, with a most likely value of 3°C (5.4°F).

In Gilbert Plass's day, the net feedback effect of clouds on global warming was also a mystery, and it's still an open question more than 50 years later. Climate contrarians seize on this uncertainty to argue that clouds will act as a big dampening effect and save us from dangerous global warming. For example, Richard Lindzen is one of the few climate scientists who is unconcerned about climate change, because he believes clouds will act as a large negative feedback, dampening global warming (more about Lindzen in chapters 3 and 6).

The first problem for climate contrarians is that there are very few known feedbacks that will significantly dampen climate change, whereas we know that water vapor and melting ice will significantly amplify global warming. Clouds are probably their best shot.

The second problem for climate contrarians is that there's no evidence that clouds act as a significant dampening feedback. As discussed earlier, all evidence (past climate change records, climate models, recent observational data, etc.) paint a consistent picture that the net effect of feedbacks will significantly amplify the global warming caused by the increasing greenhouse effect. Research into the cloud feedback specifically, for example, by the aforementioned Andrew Dessler, suggests that clouds act as a weak positive feedback in the short term, although their long-term effect is still an open question.[17]

On top of that, the fact that a variety of different methods all arrive at approximately the same estimated range of the overall sensitivity of the climate is yet another problem for contrarians. If clouds or some other effect did act to significantly dampen global warming in reality, we would see it in the geologic record, or in climate models, or in the other methods used to estimate climate sensitivity. The energy balance estimates were their last best chance, but now they appear to be in agreement with the other methods of estimating climate sensitivity, and those estimates indicate that the Earth's overall sensitivity to the increased greenhouse effect is relatively high.

3

The Astounding Accuracy of Early Climate Models

The 1960s and 1970s were a time of rapid growth in the field of climate science. Improvements in computing technology and our understanding of the climate allowed scientists to create simple climate models to predict how the Earth would respond to increases in atmospheric carbon dioxide and other climate influences.

Technically, any representation of the Earth's climate is a climate model. These can range from a simple equation to a complex mathematical and physical representation of the many different interactions in the global climate. The better we understand how the climate operates, the more complexity we can put into a model, which will make its representation of the Earth's climate more and more accurate.

In the 1970s, climate scientists understood some of the basic inner workings of the climate: for example, that increasing greenhouse gases in the atmosphere would cause warming and that the Earth's oceans and atmosphere interact with each other, exchanging heat. However, there were a lot of aspects about how the climate works that scientists did not yet understand. In addition, computer technology was still relatively primitive, so the models developed at the time were very basic representations of the climate (as opposed to today's climate models, which are extremely complex and run on some of the world's fastest supercomputers).[1]

One of the major challenges of the 1970s was to determine whether dueling warming or cooling climate effects would win out. In the wake of World War II, economies were growing fast and human fossil fuel consumption had accelerated, with carbon dioxide emissions along with it. However, sulfur dioxide is another by-product of fossil fuel combustion, which creates sulfate aerosols. These are small particulates that remain suspended in the atmosphere for about a year or two after they're first released. Sulfate aerosols have a direct cooling effect on the planet because they scatter incoming sunlight, and they also have an indirect cooling effect by seeding clouds, which in turn reflect sunlight when they form.

Leading up to the 1970s, both human carbon dioxide and sulfur dioxide emissions were rapidly accelerating. Between 1940 and 1970, there was little change in the Earth's average surface temperature, as the carbon dioxide warming and sulfur dioxide cooling roughly canceled each other out. The challenge for climate scientists was in determining which of these dueling effects would win out in the long run.

1971

In 1971, climate scientists S. Ichtiaque Rasool and Stephen Schneider published a famous paper predicting that if the amount of sulfate aerosols in the atmosphere quadrupled, it would be sufficient to cause a global cooling of up to 3.5°C (6.3°F).[2] This amount of global cooling would be sufficient to effectively trigger a new ice age. *Time* magazine in 1974 and *Newsweek* magazine in 1975 ran stories based on Rasool and Schneider's work and similar studies, painting a bleak picture about impending global cooling and a potential new ice age.[3] Many climate contrarians have seized upon these studies and magazine articles to argue that since climate scientists supposedly got their global cooling predictions wrong in the 1970s, they could be getting their global warming predictions wrong now. It's the popular "they were predicting global cooling in the 1970s" myth.

There are two major flaws in this climate contrarian argument. First, a scientific literature review between 1965 and 1979, performed by Thomas Peterson of the National Oceanic and Atmospheric Administration and colleagues, found that over 60 percent of studies during this time predicted future global warming, whereas only 10 percent predicted global cooling (the remainder did not make predictions one way or the other).[4] So in reality, climate scientists in the 1960s and 1970s were predicting global warming by a wide margin (Figure 3.1).

The second flaw in this argument is the failure to consider why the global cooling scenarios were not realized. For example, Rasool and Schneider were very clear in predicting global cooling *if* human sulfur dioxide emissions continued to accelerate, as they were leading up to the 1970s.

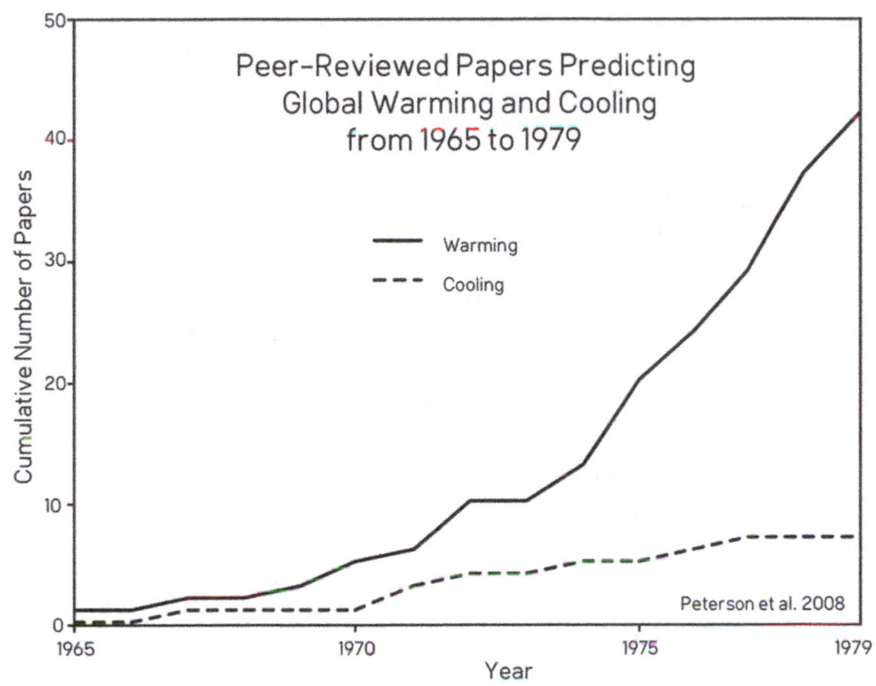

Figure 3.1 Peer-Reviewed Climate Science Papers Predicting Warming and Cooling from 1965 to 1979

This is technically called a "projection" and not a "prediction." A projection predicts what will happen in a specific scenario; if the scenario isn't realized, that doesn't make the projection wrong; it means the projection isn't applicable. For example, if I say "I'll gain 10 pounds if I go on an all-ice cream diet," but I don't start eating ice cream at every meal, that doesn't mean I've made an incorrect prediction if I don't gain 10 pounds.

That was the case for the global cooling projections. Sulfur dioxide also causes acid rain and public health problems, which were growing concerns during the late 1960s and early 1970s. As a result, a number of countries passed or amended Clean Air Acts to address the human and environmental health concerns associated with the rising sulfur aerosol pollution problem. The United States, for example, passed an amendment to its Clean Air Act in 1970 to get this pollution under control.

As a result of these international actions, sulfur dioxide pollution, which had been growing rapidly up until 1970, flattened out. Global sulfur dioxide emissions peaked in the late 1970s and have declined since then, although they briefly increased in recent years due to the rapid growth in Chinese coal consumption. However, the Chinese government has begun taking measures to control sulfur dioxide emissions to clean its air, and global sulfur dioxide emissions currently remain below 1970s levels.

In short, not only did Rasool and Schneider's scenario of quadrupling atmospheric sulfur dioxide fail to materialize, atmospheric levels actually *declined*.[5]

Thus, even for the few 1970s projections of possible global cooling, it's not accurate to say those projections were incorrect. The scenario on which their predictions were based simply was not realized because we took action to avoid it. Similarly, if we manage to reduce global greenhouse gas emissions, we can avoid possible scenarios involving catastrophic global warming.

However, many climate contrarians are forgetting history and only remember the *Time* and *Newsweek* articles about global cooling. The George Santayana quote once again comes to mind:

> Those who cannot remember the past are condemned to repeat it.[6]

1972

John Stanley (J.S.) Sawyer was a British meteorologist born in 1916. He was elected a Fellow of the Royal Society in 1962 and was also a Fellow of the Meteorological Society and the organization's president from 1963 to 1965.

A paper authored by Sawyer and published in the journal *Nature* in 1972 reveals how much we knew about the fundamental workings of the global climate over 40 years ago.[7] For example, Sawyer addressed the myth and misunderstanding that as a trace gas in the atmosphere, it may seem natural to assume that rising levels of carbon dioxide don't have much impact on the climate. Sawyer wrote:

> Nevertheless, there are certain minor constituents of the atmosphere which have a particularly significant effect in determining the world climate. They do this by their influence on the transmission of heat through the atmosphere by radiation. Carbon dioxide, water vapour and ozone all play such a role, and the quantities of these substances are not so much greater than the products of human endeavour that the possibilities of man-made influences may be dismissed out of hand.

Sawyer referenced the work by Guy Callendar in the late 1930s and early 1940s, in which Callendar estimated that the amount of carbon dioxide in the atmosphere had increased by about 10 percent over the prior 100 years. Sawyer also referenced "the Keeling Curve," which included continuous reliable measurements of the amount of carbon dioxide in the atmosphere beginning in 1958. Compared to measurements of human carbon dioxide emissions from burning fossil fuels, Sawyer noted that only about half of those human emissions were remaining in the atmosphere. The other half, climate scientists had concluded, was being absorbed by the oceans and the biosphere. Sawyer wrote:

> Industrial development has recently been proceeding at an increasing rate so that the output of man-made carbon dioxide has been increasing more or less exponentially. So long as the carbon dioxide output continues to increase exponentially, it is reasonable to assume that about the same proportion as at present (about half) will remain in the atmosphere and about the same amount will go into the other reservoirs.

Indeed, over the past four decades, human carbon dioxide emissions have continued to increase more or less exponentially, and about half has continued to remain in the atmosphere with the other half accumulating in natural reservoirs. The carbon dioxide being absorbed by the oceans has contributed to the problem of ocean acidification, sometimes referred to as "global warming's evil twin" because it's on track to cause major disruptions of marine ecosystems, particularly the widespread mortality of coral reefs.

Sawyer's paper also showed that climate scientists in the early 1970s had a good idea how quickly carbon dioxide levels in the atmosphere would continue to rise as a result of human activities.

> Bolin has estimated that the concentration of carbon dioxide will be about 400 ppm by the year 2000. A recent conference put the figure somewhat lower (375 ppm).

That prediction at the referenced 1971 conference on "the Study of Man's Impact on Climate"[8] turned out to be quite accurate. In 2000, atmospheric carbon dioxide concentrations were measured at about 370 ppm. Fifteen years later, they're now right around 400 ppm.

In his paper, Sawyer discusses the predicted impacts resulting from a continued rise in atmospheric carbon dioxide. He noted that directly "it might make some vegetation grow a little faster," which is generally true, although the situation is complicated because the climate change associated with that rising carbon dioxide tends to cause more extreme weather like heat waves and floods that are generally bad for plant growth.

Sawyer noted that rising carbon dioxide levels would cause an increased greenhouse effect, and the associated warming would lead to more evaporation and more water vapor in the atmosphere. As a greenhouse gas itself, that rise in water vapor would act to further amplify human-caused global warming.

> If world temperatures rise due to an increase in carbon dioxide, it is almost certain that there will be more evaporation of water—the water vapour content of the atmosphere will also increase and will have its own effect on the radiation balance.

Sawyer referenced a 1967 paper by Manabe and Wetherald of the Environmental Science Services Administration, in which the scientists had calculated that a doubling of atmospheric carbon dioxide would by itself cause approximately 1.3°C (2.3°F) global surface warming, but that warming would be amplified by a further 1.1°C (2.0°F) due to rising water vapor concentrations if the relative humidity were to remain constant.[9] Observations have indeed unequivocally shown that rising levels of water vapor in the atmosphere strongly amplify human-caused global warming, for example, as found in a 2009 paper by Andrew Dessler and Sun Wong from Texas A&M University.[10]

Sawyer put all this information together to predict how much average global surface temperatures would warm between 1972 and 2000.

> The increase of 25% CO_2 expected by the end of the century therefore corresponds to an increase of 0.6°C in the world temperature—an amount somewhat greater than the climatic variation of recent centuries.[11]

Remarkably, between 1850 and 2000, atmospheric carbon dioxide levels did increase by very close to 25 percent, and global average surface temperatures also increased by just about 0.6°C (1.1°F) during that time.

Sawyer also discussed that melting ice and snow in a warming world would act to amplify global warming, but suggested that increasing cloud cover might dampen global warming and act to regulate the global climate. However, as previously discussed, recent research by Andrew Dessler and others has shown that clouds may actually weakly amplify global warming as well.

Sawyer's paper noted that climate scientists in the early 1970s understood that it would take on the order of 100 years (now understood to be even longer) for the planet to reach a new energy equilibrium from the energy imbalance caused by the increased greenhouse effect. This lag is due to the thermal inertia of the oceans; the water in the oceans stores heat for a long time, but ultimately the global oceans and atmosphere interact with each other until a new energy balance is reached. Sawyer also understood that significant global warming would cause changes in weather and wind patterns around the world.

At the same time, Sawyer got a bit lucky. His prediction did not account for natural influences on global temperatures, like changes in solar and volcanic activities, or for natural ocean cycles, or for other human influences on the climate besides carbon dioxide like sulfur aerosols. However, as it turns out, the influences of these other factors have approximately canceled each other out since the 1970s. Human emissions of greenhouse gases other than carbon dioxide have caused some warming, but human emissions of aerosol pollution have caused some cooling. Solar activity has been very slightly downward, and volcanic emissions have been fairly steady as well. Ocean cycles have gone up and down, also having little net effect on global surface temperatures since the 1970s.

While Sawyer was fortunate that these influences have had very little net effect on global surface temperatures, at the same time, he was smart to realize that human carbon dioxide emissions would be a dominant driver of global warming.

All in all, Sawyer's 1972 paper demonstrated a solid understanding of the fundamental workings of the global climate and included a remarkably accurate prediction of global warming over the next 30 years. This goes to show that climate scientists have understood the main climate control knobs for over four decades.

1975

American climate scientist Wallace Broecker was among the first scientists to use simple climate models to predict specific future global temperature changes. His 1975 paper *Climatic Change: Are We on the Brink of a Pronounced Global Warming?*[12] is also widely credited with coining the term "global warming."

In that paper, Broecker modeled the effects of the expected future increase of carbon dioxide due to humans burning fossil fuels, combined with a natural climate cycle which he estimated based on Greenland ice core records (which he called "Camp Century cycles"). Broecker tweaked his simple climate model to match the observed temperature record at the time, starting around 1900.

This was a very simple model, excluding the effects of the sun, volcanoes, other greenhouse gases, sulfate aerosols, and so forth, which Broecker acknowledged:

> In this report only the interaction of the CO_2 effect and natural climatic change is considered. As other anthropogenic effects are shown to be significant and as means to quantitatively predict their future influence on global temperatures are developed, they can be included in models such as this.

Technically, by making his model match the observed global temperature from 1900 to 1975, unlike Sawyer before him, Broecker did include natural effects like the sun and volcanoes in his approximation of natural climate cycles. However, this is an accurate approach only if all natural effects are cyclical, which is not always the case. For example, although the sun has natural cycles, the amount of solar radiation reaching the Earth's surface over the long term is unpredictable. Climate models can include the effects of past solar changes on global temperature, but they can't predict future solar changes.

However, solar activity tends to be very stable, especially compared to the size of current carbon dioxide increases and their influence on global temperatures. For example, solar research has suggested that we could be due for another "grand solar minimum" within the next century. There have been two grand solar minima in the past 500 years—the Maunder Minimum between 1645 and 1715 and the Dalton Minimum between 1790 and 1830. These two periods of low solar activity contributed to an event known as the "Little Ice Age," during which there was some global cooling, but the cool temperatures were primarily localized in Europe.

Despite its name, the Little Ice Age was not very cold across the Earth as a whole. In 2013, a team of 78 researchers from 60 separate scientific institutions around the world (part of the Past Global Changes, or PAGES Network) collaborated to create the most comprehensive reconstruction of global surface temperatures over the past 2,000 years.[13] This collaboration allowed experts on local temperature reconstructions from each continent to contribute their expertise to create the best estimate to date of temperature changes over the past two millennia.

The resulting PAGES reconstruction estimated that between the time of hottest temperatures during what's known as the "Medieval Warm Period" about 1,000 years ago to the coldest temperatures during the Little Ice Age, there was only about a half degree Celsius cooling (less than 1°F). From that coolest part of the Little Ice Age to today, including recent instrumental temperature measurements by thermometers, we've seen double that—over 1°C (about 2°F) global surface warming. Current global surface temperatures are likely the hottest they've been in thousands of years and quite possibly the hottest over the past tens of thousands of years.

Nevertheless, there have been several papers published in the past few years to determine just how much cooling another grand solar minimum could cause. These papers have estimated that during the Maunder and Dalton Minima, the amount of solar radiation reaching the Earth's surface declined by 0.25 percent and 0.08 percent, respectively. That's how stable solar activity is—even during a grand solar minimum, the solar energy reaching Earth drops only by a fraction of a percent.

This is fortunate for the species on Earth, because it results in a very stable climate, which has allowed human civilization to thrive. We've been able to create large stationary agricultural farms because the stable climate over the past 10,000 years allowed us to rely on stable weather patterns. Unfortunately, we're now in the process of destabilizing the climate by rapidly increasing the greenhouse effect and throwing the Earth's energy balance out of whack. We're running a very dangerous experiment; we don't know how it will turn out, but all signs point to dangerous consequences.

Scientists from the Potsdam Institute for Climate Impact Research in Germany,[14] the Met Office Hadley Centre in the United Kingdom,[15] the Institute for Atmospheric and Climate Science in Switzerland,[16] the National Center for Atmospheric Research (NCAR) in Colorado,[17] and the NASA Goddard Institute for Space Studies (GISS)[18] are among those investigating the climate consequences of a possible new grand solar minimum by using global climate models. These papers have all arrived at remarkably similar conclusions—namely that a new grand solar minimum would cause no more than 0.3°C (0.54°F) global surface cooling. This would cause only a small dent in the several degrees of global surface warming that will occur over the next century, which is likely to be several degrees. Thus, a grand solar minimum would offset just a few years' worth of global warming due to human greenhouse gas emissions.

Moreover, any cooling from a grand solar minimum would be temporary, lasting only until the sun once again became more active. A solar minimum would only last for a few decades, and once it were to end, the associated cooling would be offset by solar warming once again. While the Little Ice Age lasted longer than a few decades, the sun was only one contributing factor causing cooling at that time. Recent research has concluded that a period of heightened volcanic activity and changes in atmospheric greenhouse gas levels probably played a bigger role in the Little Ice Age cooling than solar activity.[19]

In his 1975 research, Wallace Broecker couldn't predict how solar activity would change in the future, nor was he able to account for the cooling influence of human sulfate aerosol emissions or the warming influence from greenhouse gases other than carbon dioxide. As it turns out, like Sawyer before him, Broecker was fortunate that the cooling effects of human sulfate aerosols have roughly canceled out the warming effects of human non-carbon greenhouse gases (e.g., methane) since 1975, and solar and volcanic activities have been relatively flat over that period. So the net effect of the factors that he did not take into account has been close to zero. However, like Sawyer, Broecker was also smart; the dominant effect on temperature since 1975 has been from carbon dioxide, as he expected. It's better to be lucky than good, but it's best to be both.

To create a future global temperature prediction, Broecker also had to predict how human carbon dioxide emissions would change in the decades ahead. Broecker anticipated the actual increase in atmospheric carbon dioxide very closely, predicting 373 ppm in 2000 and 403 ppm in 2010 (the actual values were 369 and 390 ppm, respectively). Broecker also used a model with an equilibrium climate sensitivity of 3°C (5.4°F) global surface warming in response to a doubling of atmospheric carbon dioxide.

However, it takes time for the global climate to reach a new energy equilibrium in response to an energy imbalance created by a radiative forcing like an increased greenhouse effect. This is mainly due to the aforementioned thermal inertia of the world's oceans. The vast majority of a global energy imbalance goes into heating the oceans, but it doesn't stay there forever, because the Earth's oceans and air interact with each other. It takes time, but eventually the planet's surface reaches the amount of warming it's committed to by a given global energy imbalance.

In the meantime, approximately two-thirds of the eventual equilibrium surface warming will be realized almost immediately. However, Broecker's model overestimated how much global warming would be immediately realized at the Earth's surface. As a result, in the short term, his model was roughly equivalent to a 3.6°C (6.5°F) equilibrium climate sensitivity for doubled atmospheric carbon dioxide, which is within the range of today's climate models (2.0 to 4.5°C, 3.6 to 8.1°F), but a bit higher than the current best estimate of 3°C (5.4°F).

So how accurate was Broecker's projection of future warming? Figure 3.2 uses Broecker's model, taking into account the actual atmospheric carbon dioxide changes since 1975 (which, as noted earlier, have been slightly lower than what Broecker anticipated) and compares the results to the best available measured global (land and ocean) surface temperature change estimates, generated by my colleagues Kevin Cowtan and Robert Way.

Broecker's overestimate of the global surface warming between 1900 and 1940 reveals that global surface temperature data sets were not as accurate in 1975 as they are today. However, Broecker's prediction has matched the net global temperature change quite closely since 1975.

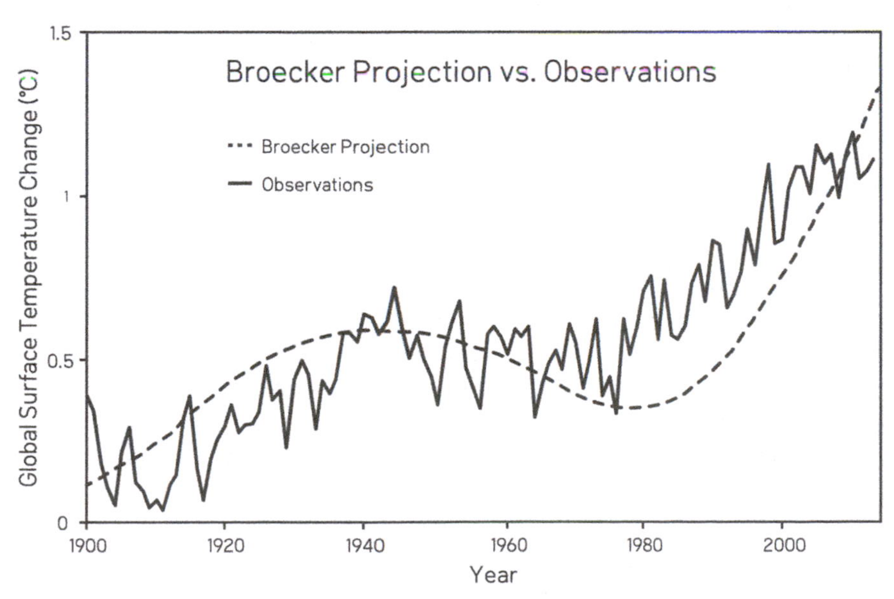

Figure 3.2 Broecker's 1975 Climate Model, Accounting for Actual Carbon Dioxide Changes versus Observed Temperature

Broecker's "natural cycle" predicted a natural cooling effect from about 1970 to 2005 and thus held his overall temperature prediction below the actual global temperature increase for most of the period. Broecker's model predicted that the natural cycle's influence on global temperatures would approach zero after 2000. Not coincidentally, this is when Broecker's prediction most closely matches the observed global warming.

Broecker somewhat overestimated the amount of global warming between 1975 and 2013, by about 0.3°C (0.54°F). This is probably mainly due to his slight overestimate of the Earth's climate sensitivity. It's quite remarkable that a prediction made in 1975 using such a simple model of the climate could so accurately match the observed global temperature change. Once again, it's a testament to the dominant effect of carbon dioxide and the fact that we have had a solid understanding of the fundamental workings of the Earth's climate for many decades.

1979

William Kellogg was an American meteorologist and climate scientist who served as associate director and senior scientist at NCAR and held a number of prestigious scientific positions in his career, including serving on the President's Science Advisory Committee and the National Academy of Science's Space Science Board.

In 1979, Kellogg authored an extensive review paper summarizing the state of climate modeling at the time.[20] Among the studies referenced in Kellogg's work was Wallace Broecker's 1975 study.

Kellogg's review discussed the fact that in the late 1970s, climate models were still relatively simple and excluded or did not accurately reflect some important climate feedbacks (such as cloud cover changes); however, they did include the feedback from changes in the reflectivity of the Earth's surface due to retreating or advancing ice in response to changing temperatures. As previously discussed, when ice melts, it reveals the much darker ground or ocean surface below; this decreased reflectivity causes the Earth to absorb more solar radiation and warm even further and is thus a positive feedback that amplifies global warming (or cooling).

As in Broecker's 1975 study, Kellogg correctly identified that carbon dioxide represents the most significant human impact on the global climate. However, Kellogg also thought that sulfate aerosols should have a net warming effect on the climate, because not only do they scatter sunlight, but they also absorb it; Kellogg believed the latter effect was stronger than the former. However, we now know that Kellogg was incorrect on this point. Based on up-to-date climate research, aerosols certainly have a net cooling effect and possibly a very strong one.

In his 1979 paper, Kellogg predicted future polar and global surface temperature changes in two scenarios involving "low" and "high" greenhouse gas emission levels. His high scenario, which involved accelerating greenhouse gas emissions, has been close to the observed emissions growth since 1979.

Strangely, Kellogg predicted that future temperatures would rise in linear (increasing at a steady rate) fashion, even though he anticipated essentially the same exponential (accelerating) atmospheric carbon dioxide increase as Broecker did in his 1975 study. It's unclear why Kellogg predicted a steady increase in global temperature rather than an accelerating increase like Broecker's projection. Regardless, we can compare Kellogg's high projection to the observed global temperature increase (Figure 3.3).

Clearly, Kellogg significantly overestimated the ensuing global warming—much more so than Broecker a few years earlier. So what went wrong? In his paper, Kellogg said that most models were similarly sensitive to changes in atmospheric carbon dioxide as Broecker's model.

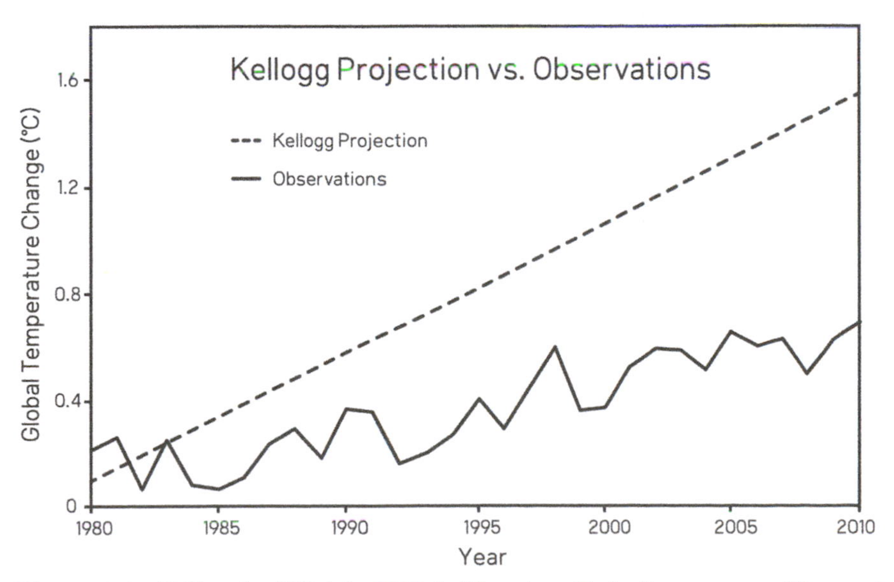

Figure 3.3 Kellogg's "High" Global Warming Projection versus Observed Temperature

> The best estimate of the "greenhouse effect" due to a doubling of carbon dioxide lies between 2 and 3.5°C increase in average surface temperature . . . and both the models and the record of the behavior of the real climate show that the change in the polar regions will be greater than this by a factor of from 3 to 5, especially in winter.[21]

This climate sensitivity is also broadly consistent with that anticipated by today's climate models of 2 to 4.5°C (3.6 to 8.1°F) global surface warming in response to a doubling of atmospheric carbon dioxide. However, Kellogg's temperature prediction used a model with higher sensitivity than the range mentioned earlier:

> When the level of carbon dioxide has risen to 400 ppmv from its present 330 ppmv, the rise in average surface temperature is estimated to be about 1°C. These figures refer to the effect of carbon dioxide alone.[22]

An average global temperature response of 1°C (1.8°F) to a carbon dioxide increase of 330 to 400 ppm corresponds to a climate sensitivity equivalent to 3.6°C (6.5°F) surface warming in response to a doubling of atmospheric carbon dioxide. However, it appears that Kellogg assumed that this temperature response would be instantaneous, with the average global temperature warming 1°C by the time atmospheric carbon dioxide levels reached 400 ppm (by the year 2011, in his estimation).

As previously noted, the instantaneous temperature response is approximately two-thirds as large as the total eventual warming once the planet reaches a new energy equilibrium. Therefore, the equilibrium climate sensitivity employed in Kellogg's prediction is equivalent to approximately 5.4°C (9.7°F), which is quite high, and above the current accepted range of equilibrium climate sensitivity values. This discrepancy explains much of his overestimated global warming. Kellogg neglected the thermal inertia of the oceans.

Kellogg also included the warming effects of other greenhouse gases (such as methane) in his model, but did not include the cooling effects of sulfate aerosols (which, as noted earlier, Kellogg believed had a net warming effect as well, but which we now know have a significant cooling effect). By including the warming effects of other greenhouse gases, the equivalent carbon dioxide concentration in Kellogg's model reached 400 ppm by 2000, causing him to overestimate the rate of warming even further.

It's worth noting that because Kellogg's prediction was linear while the actual temperature increase will accelerate, the further ahead in time we go, the more accurate Kellogg's prediction will become. In 2050 it will be less inaccurate than in 2015, but due to the high sensitivity of his model and the other incorrect assumptions he used, the actual temperature will still be below Kellogg's prediction.

However, Kellogg was quite accurate in one aspect of his prediction: polar amplification. The planet warms fastest at the poles because of the local feedbacks, the biggest of which is the aforementioned decrease of reflectivity due to melting ice. This is a particularly large temperature-amplifying influence in the Arctic; when highly reflective sea ice melts and reveals the dark oceans below, more solar radiation is absorbed, causing additional warming.

The average Arctic surface temperature has increased approximately 3°C (5.4°F) since 1880 and approximately 2°C (3.6°F) since 1970. This is a warming rate approximately 3.6 times faster than the average global surface warming, which is within Kellogg's predicted polar amplification range of a factor of 3 to 5. Thus, although his global warming prediction was inaccurate, there are aspects of Kellogg's work that we can learn from.

1981

James Hansen is an American climate scientist at Columbia University and NASA GISS (now retired), who is considered one of the world's foremost climate scientists. In 1981 (the year I was born), Hansen published a paper projecting future global surface temperature changes.[23] Like Broecker and Kellogg before him, by 1981 Hansen already realized the important role carbon dioxide was playing in global warming, writing,

> The global temperature rose by 0.2°C between the middle 1960's and 1980, yielding a warming of 0.4°C in the past century. This temperature increase is consistent with the calculated greenhouse effect due to measured increases of atmospheric carbon dioxide. . . . It is shown that the anthropogenic carbon dioxide warming should emerge from the noise level of natural climate variability by the end of the century, and there is a high probability of warming in the 1980's.

Hansen was correct to predict warming during the rest of the 1980s; there was indeed nearly 0.1°C (0.17°F) global surface warming between

1981 and 1990. Hansen and his colleagues also noted that at the time, the human-caused global warming theory was having difficulty gaining traction because of the flat mid-century surface temperatures.

> The major difficulty in accepting this theory has been the absence of observed warming coincident with the historic CO_2 increase.

This is an argument still made by climate contrarians today to assert that greenhouse gases aren't driving global warming, because carbon dioxide was rising during the mid-20th century while temperatures remained flat.

However, this argument is flawed because greenhouse gases aren't the only factors that influence global surface temperatures. After World War II, fossil fuel consumption rose rapidly. As previously discussed, this caused carbon dioxide emissions to rise, but also sulfur aerosol emissions and their associated cooling effect, caused by deflecting sunlight. Only in the 1970s when Clean Air Acts were implemented to reduce sulfur dioxide emissions in order to address the problem of acid rain did their cooling influence begin to dissipate, and the warming influence of carbon dioxide then took over.

Hansen recognized this shift was occurring and thus correctly predicted that as long as human greenhouse gas emissions continued to rise, the planet would continue to warm. Despite being relatively primitive, in the early 1980s, the sensitivity of climate models to the increasing greenhouse effect was very similar to that of today's much more complex models. As Hansen wrote in his 1981 paper,

> The most sophisticated models suggest a mean warming of 2° to 3.5°C for doubling of the CO_2 concentration from 300 to 600 ppm.

This is very close to today's best estimate of 3°C (5.4°F) global surface warming in response to a doubling of atmospheric carbon dioxide, with a likely range between 1.5 and 4.5°C (2.7 to 8.1°F). In order to project how much the Earth's surface would warm in the future, Hansen used a climate model whose sensitivity was 2.8°C (5.0°F) for a doubling of atmospheric carbon dioxide.

Hansen and his colleagues ran their global climate model using combinations of the three main effects on global temperatures (carbon dioxide, the sun, and particulates released from volcanoes). They projected how much global surface temperatures would warm based on various energy growth scenarios. They also modeled various scenarios involving fossil fuel energy source replacement starting in 2000 and in 2020.

Between 1981 and today, actual greenhouse gas emissions have risen between the "slow" and "fast" scenarios considered in Hansen's study. As Figure 3.4 shows, the model accuracy has been very impressive.

Hansen's model projected 0.17°C (0.31°F) global surface warming per decade in the fast scenario and 0.13°C (0.23°F) per decade in the slow scenario. Thus, based on actual greenhouse gas emissions since 1981 falling between the two scenarios, we would expect to see about 0.15°C (0.27°F) surface warming per decade.

In actuality, we've observed about 0.18°C (0.32°F) warming per decade, about 20 percent higher than Hansen projected. This suggests that Hansen's 1981 global climate model was probably a bit less sensitive to the increasing greenhouse effect than the real world, at least in the short term.

Because James Hansen is such a prominent world-renowned climate scientist and also a strong advocate for taking action to address the threat posed by climate change, climate contrarians often try to label him as an "alarmist." Seven years later, Hansen updated his global surface warming projections with a newer climate model that was

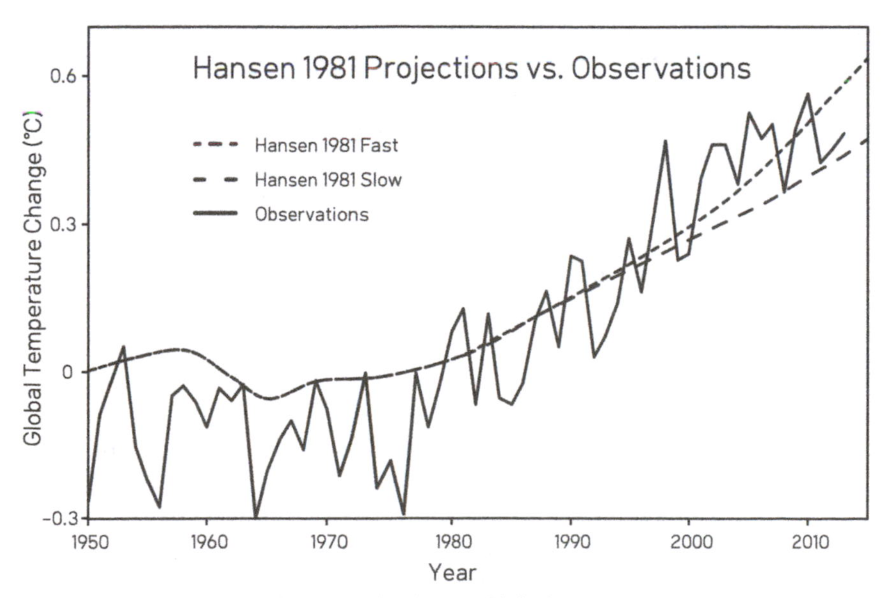

Figure 3.4 Hansen's "Slow" and "Fast" Global Warming Projections versus Observed Temperature

more sensitive to the increased greenhouse effect, and at that time his model did over-project global surface warming. Not surprisingly, climate contrarians focus on Hansen's 1988 model projections. However, in 1981, Hansen's climate model actually somewhat underestimated global surface warming.

1988

In 1988, Hansen published what has become a very well-known paper in which he used a global climate model to simulate the impact of changes in atmospheric greenhouse gases and sulfate aerosols from volcanic eruptions on the global climate.[24] Hansen chose three scenarios to model for his global warming projections. His Scenario A assumed continued exponential (accelerating) greenhouse gas growth, Scenario B assumed a linear rate of growth, and Scenario C assumed a rapid decline in human greenhouse gas emissions around 2000.

Of the three scenarios, thus far, Scenarios B and C have been closest to reality. Although we didn't take the dramatic steps Hansen's Scenario C envisioned to significantly reduce global greenhouse gas emissions by 2000, we did implement the Montreal Protocol. This was an international agreement to phase out the use of chlorofluorocarbons (CFCs) because they were causing ozone depletion and created a hole in the ozone layer. However, CFCs are also greenhouse gases, so phasing out their use in industrial products also helped slow the rise of the greenhouse effect. Fortunately, Hansen's Scenario A with its rapidly accelerating greenhouse gas emissions has turned out to be the least realistic.

The sensitivity to increasing carbon dioxide in Hansen's 1988 model was a bit on the high side as well, with an average global surface warming of 4.2°C (7.6°F) in response to a doubling of atmospheric carbon dioxide. Since the current best estimate of equilibrium climate sensitivity is 3°C (5.4°F), when comparing his projections to the observed temperatures, we should expect them to somewhat overestimate global warming.

As a result, Figure 3.5 compares Hansen's Scenario B, adjusted to reflect the actual greenhouse gas levels and other energy imbalance changes from 1988 to 2013 into account, with the observed global surface temperature change.

As expected, due to its high climate sensitivity, Hansen's model projected a bit more warming than was observed by the end of 2013. Hansen's projected surface warming trend between 1988 and 2013 was 0.25°C (0.45°F) per decade, whereas the observed trend has been

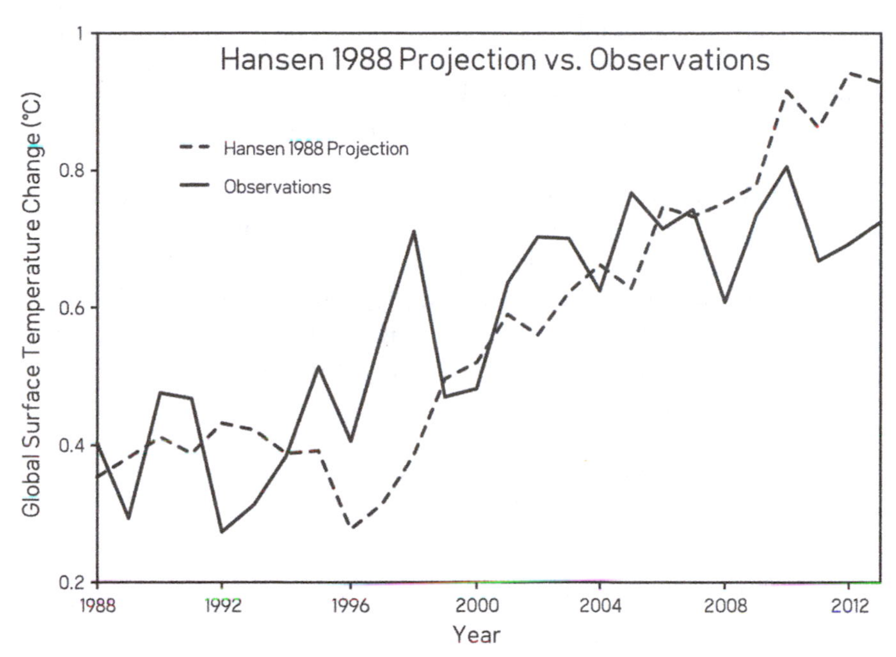

Figure 3.5 Hansen's Scenario B, Adjusted to Reflect Actual Global Energy Imbalance Changes since 1988 versus Observed Temperature

0.18°C (0.32°F) per decade. We might then ask the question, If Hansen had access to today's global climate models in 1988, how accurately would they have predicted the ensuing global warming?

Today's climate models have an average equilibrium climate sensitivity of around 3°C (5.4°F), about 30 percent lower than that in Hansen's 1988 model. Likewise, Hansen's model projected about 30 percent more global surface warming than has actually been observed over the past 25 years. Hence, if Hansen had access to today's climate models in 1988, his global warming projections would have been very close to what we've seen in recent decades. Combined with Broecker's and Kellogg's results, this result provides additional supporting evidence for the real-world equilibrium climate sensitivity estimate of 3°C global surface warming if the amount of carbon dioxide in the atmosphere doubles.

In 1988, Hansen also testified before U.S. Congress. He presented the results of his global warming projections and made three main points:

> Number one, the earth is warmer in 1988 than at any time in the history of instrumental measurements. Number two, the global warming is now large enough that we can ascribe with a high degree of

> confidence a cause and effect relationship to the greenhouse effect. And number three, our computer climate simulations indicate that the greenhouse effect is already large enough to begin to affect the probability of extreme events such as summer heat waves.[25]

Climate contrarian Patrick Michaels of the conservative think tank and political advocacy group Cato Institute (who in an August 2010 interview on CNN admitted that approximately 40 percent of his funding comes from the petroleum industry[26]) was invited to testify before Congress on July 29, 1998.[27] During that testimony, Michaels grossly misrepresented Hansen's global warming projections from 10 years earlier, claiming:

> Ground-based temperatures from the IPCC show a rise of 0.11°C, or more than four times less than Hansen predicted. . . . The forecast made in 1988 was an astounding failure.

Figure 3.5 clearly shows that Hansen's projections were not off by anywhere near a factor of four, so how can Michaels possibly justify this claim?

In the graphic he presented to Congress to illustrate Hansen's 1988 global warming projections, Michaels presented only Hansen's Scenario A, even though this was the scenario furthest from reality. The global temperature change from 1988 to 1998 was very close to the Scenario B projection. Not coincidentally, Scenarios B and C were also the most representative of actual emissions and the global energy imbalance. By erasing the most accurate scenarios in Hansen's study and presenting the least representative scenario as his "prediction," Michaels committed borderline perjury and certainly misinformed our policymakers in the process.

This misrepresentation was also reproduced in Michael Crichton's science fiction novel *State of Fear*, whose plot is built around a group of ecoterrorists attempting to create widespread fear and panic to further advance their "global warming agenda." The novel featured a scientist claiming that Hansen's 1988 projections were "overestimated by 300 percent."[28] This was just one of many scientific errors and misrepresentations in the novel, as much as this pains me as a fan of much of Michael Crichton's science fiction writing. Unfortunately, *State of Fear* was heavy on fiction and light on accurate science.

These examples illustrate the lengths to which climate contrarians will go in order to make their anti-climate case and misinform the public in the process. High-profile climate scientists like James Hansen continue to be frequent targets of fossil fuel–funded conservative think tanks and climate contrarian blogs, despite (or perhaps because of) the fact that their global warming predictions have turned out to be quite accurate.

1989

Thus far, we have examined only global temperature projections made by mainstream climate scientists who agree that the global warming over the past century is primarily man-made, because very few climate contrarians have made similar predictions.

There are various levels of climate contrarians. Some reject very well-established science, like the fact that the planet is even warming. Others accept that the planet is warming but don't believe human greenhouse gas emissions are the main cause. These contrarians argue that because they believe humans aren't the main cause of the current global warming, future global warming will be relatively small, and we'll be able to adapt to its consequences without the need to prevent them by reducing greenhouse gas emissions.

Most climate scientists are real skeptics, in the sense that a skeptic keeps an open mind and examines all new evidence with a critical eye. The reason I don't refer to contrarians as "skeptics" in this book is that most self-proclaimed climate change skeptics are not really skeptical at all. The majority of people in this group seek information to support what they want to believe—that humans are not causing dangerous global warming. Rejecting information that does not fit one's predetermined conclusions is not skepticism; it's biased closed-mindedness. It's fine to be skeptical and ask questions, but rejecting the body of available evidence because it's inconvenient is not real skepticism.

In 2014, my colleague John Cook came up with "the quantum theory of climate denial," which he described as follows:

> There are various states of climate denial, with some states contradicting others. For example, some believe global warming is not happening. Others believe global warming is happening but is not caused by humans. Others believe humans are causing global warming but that the impacts won't be bad.
>
> Now, it's perfectly understandable for a community of people to hold mutually inconsistent beliefs. But can one person hold three inconsistent beliefs at the same time? Can a person argue that global warming is not happening, then smoothly transition to arguing that global warming is happening but is caused by something else?

> They can, and they do. . . . It can be explained by the "quantum theory of climate denial." This theory holds that climate deniers exist in a fuzzy quantum state of denial, simultaneously rejecting many or all aspects of climate science.[29]

The reason climate contrarians can deploy any and all of these contradictory arguments is that their denial isn't based on science or fact; rather, it's based on ideology. They want to maintain the status quo and/or oppose the proposed solutions to global warming, like renewable energy and electric vehicles, but mainly government climate policies. Since they reject the science only because of their ideological bias, the accuracy and consistency of their scientific arguments don't matter. As long as an argument can be used to justify their predetermined rejection of climate science and opposition to climate solutions, they'll use it. That mind-set, based purely on taking a position contrary to inconvenient science, is why I call them climate contrarians rather than skeptics.

Among climate scientists (those who actively conduct research to learn about the Earth's climate), numerous studies and surveys have shown that about 97 percent agree that humans are the primary cause of the global warming over the past century.[30,31,32,33] The handful of climate scientist "skeptics" agree that human greenhouse gas emissions have caused some amount of warming, but they generally believe it's a small amount because they argue climate sensitivity is much lower than scientific research has shown.

The most prominent and longest-tenured of these climate scientist "skeptics" is Richard Lindzen from the Massachusetts Institute of Technology (MIT). Lindzen retired from MIT in 2013 and soon thereafter joined Patrick Michaels at the libertarian political advocacy group and think tank Cato Institute.[34] For a climate scientist to retire from academia and join a political think tank is very unusual, but Richard Lindzen has previously argued against governments taking action to reduce greenhouse gas emissions. For example, in a testimony before a UK Parliament committee in January 2014, Lindzen claimed:[35]

> An economist at Yale, Bill Nordhaus, has a book on climate policy and if you look carefully at that book, he estimates the cost-benefit and so on of various policies and it's clear that there is virtually no policy that beats doing nothing for 50 years.

Apparently Lindzen didn't heed his own advice and look carefully at Nordhaus's book. This statement is entirely untrue, but a common misrepresentation of Professor Nordhaus's research. It's a misrepresentation so common that nearly two years earlier, Nordhaus published an article in *The New York Review of Books* refuting this argument, unambiguously titled "Why the Global Warming Skeptics Are Wrong."[36]

> My study is just one of many economic studies showing that economic efficiency would point to the need to reduce CO_2 and other greenhouse gas emissions right now, and not to wait for a half-century. Waiting is not only economically costly, but will also make the transition much more costly when it eventually takes place. Current economic studies also suggest that the most efficient policy is to raise the cost of CO_2 emissions substantially, either through cap-and-trade or carbon taxes, to provide appropriate incentives for businesses and households to move to low-carbon activities.

Perhaps Lindzen was unaware of Nordhaus's 2012 article, or the many other instances where Nordhaus and others have debunked this misrepresentation of his work. In any case, Lindzen has made it clear that he opposes governments taking action to reduce the threats posed by climate change (although his misrepresentation of Nordhaus's work makes it clear that Lindzen hasn't researched climate economics and policy in any depth). So although it was an unusual move for a retired climate scientist, it probably didn't surprise many people when Lindzen joined the Cato Institute. However, when a scientist joins a political think tank, it should perhaps make people question whether his political and ideological biases are clouding his scientific judgment.

There was a similar situation in May 2014, when a Swedish meteorologist named Lennart Bengtsson from the University of Reading in the United Kingdom briefly joined the Global Warming Policy Foundation (GWPF). The GWPF is essentially the UK version of the Cato Institute, a conservative political advocacy group periodically releasing reports full of shoddy science and then using them to lobby against climate policies.

Unlike Lindzen, Bengtsson is still actively publishing scientific research, and some of his colleagues allegedly expressed concern and withdrew from publishing further research with him. If true, these decisions would be entirely understandable, because GWPF has a reputation for attacking climate scientists and scientific organizations, and most scientists don't want any association with political groups. Bengtsson decided to withdraw his association with GWPF within a few days and then went to the conservative media with hypocritical complaints about how climate science has become politicized. What's interesting is that this now gives us two examples of the small number of contrarian climate scientists joining politically conservative advocacy groups. It's rare for those among the much larger group of mainstream climate scientists to join political organizations, although some act as advisors to environmental groups.

Richard Lindzen has never made a specific prediction as to how global temperatures will change in the future. However, in 1989, he gave a talk at MIT in which he made a number of comments relevant to global temperature change.[37] For example, Hansen and colleagues at NASA GISS began compiling a global surface temperature record called GISTEMP in 1981. As of 1988–1989, their record showed that the average global surface temperature had risen approximately 0.5 to 0.7°C (0.9 to 1.3°F) since 1880, when the record begins. Lindzen, however, disputed the accuracy of GISTEMP in his 1989 MIT talk.

> The trouble is that the earlier data suggest that one is starting at what probably was an anomalous minimum near 1880. The entire record would more likely be saying that the rise is 0.1 degree plus or minus 0.3 degree. . . . I would say, and I don't think I'm going out on a very big limb, that the data as we have it does not support a warming. Whether it contradicts it is a matter of taste.

It turns out that Lindzen's first statement here was incorrect. According to the slightly longer global surface temperature record of the UK Met Office, 1880 was closer to a local maximum global surface temperature than a minimum. But more importantly, Lindzen was claiming here that the average global surface temperature warming between 1880 and 1989 was approximately 0.1°C (0.2°F). Lindzen proceeded to effectively assert that any greenhouse gas warming signal is and will continue to be swamped out by the noise of the natural internal variability of the global climate system.

> I personally feel that the likelihood over the next century of greenhouse warming reaching magnitudes comparable to natural variability seems small.

Climate research has estimated that the natural internal variability of the climate system rarely results in more than 0.2 to 0.3°C (0.4 to 0.5°F) global surface warming or cooling over periods of several decades, so Lindzen was clearly predicting a very small amount of greenhouse warming over the next century.

Using these quotes, I reconstructed what I think is a reasonable approximation of a global surface temperature prediction based on Lindzen's belief of the small warming effects of rising greenhouse gas levels. Remember, Lindzen has not made a specific global temperature prediction; these projections are my interpretation of Lindzen's comments, not Lindzen's own projections.

By estimating the climate sensitivity assuming that Lindzen's claimed 0.1°C (0.2°F) global warming between 1880 and 1989 was totally caused by human greenhouse gas emissions, or by simply projecting a 0.1°C per century global warming trend forward from the present, either approach yields approximately the same modest global warming prediction. These are shown in Figure 3.6, with some random noise added to simulate natural climate variability. Note that Lindzen's projection begins approximately 0.4°C (0.7°F) below the observational data in 1990, reflecting Lindzen's claim that there was only 0.1°C (0.2°F) global surface warming between 1880 and 1989, when in reality the warming during that time frame was approximately 0.5°C (0.9°F).

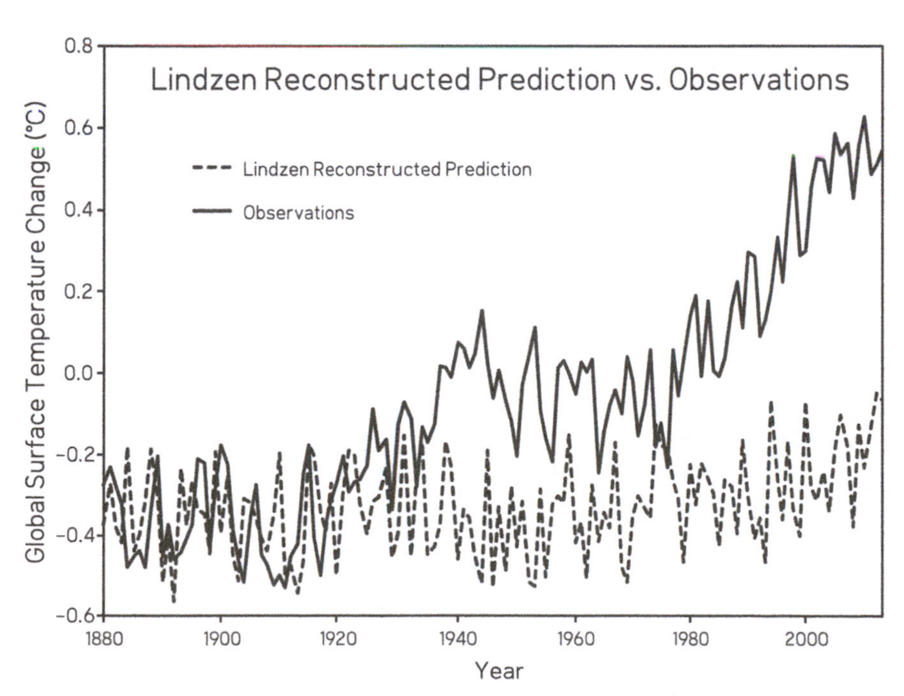

Figure 3.6 Global Temperature Projection Reconstructed from Lindzen's 1989 MIT Tech Talk Comments versus Observed Temperature

On the accuracy of the instrumental temperature record, the amount of warming between 1880 and 1989 and the amount of warming from 1989 through 2013, Hansen's mainstream climate science was fundamentally correct and Lindzen's climate "skepticism" was wrong.

In fact, Lindzen has a very long history of contrary thinking regarding nearly every facet of mainstream climate science (in addition to his "skepticism" about the strength of the link between secondhand smoke and cancer[38]). For example, in his 1989 MIT talk, Lindzen argued that the global surface temperature record was incorrect, that the Earth hadn't warmed as much as most climate scientists expected, that climate sensitivity to increasing carbon dioxide is low, and that water vapor acts to dampen global warming, among other "skeptical" positions. In every single case, subsequent research over the past two decades has proven Lindzen wrong.[39]

We now know that in reality, the NASA GISS temperature record in the late 1980s was accurate in its estimates of global surface warming since 1880 and that Lindzen was wrong to claim there was negligible warming during that time. Lindzen also claimed that according to climate models,

> Since the 19th century we should have seen between about one and two degrees of warming.

This claim was simply false, as illustrated by the climate model global warming projections shown throughout this book. None predicted one to two degrees warming (either Celsius or Fahrenheit) between 1900 and 1989. To this day, Lindzen continues to claim that climate models dramatically overpredict global warming. In fact, it's probably his favorite climate argument and one he makes on nearly an annual basis. I was able to track down similar statements made by Lindzen in[40]

- a 2002 letter to his local mayor in Newton, Massachusetts;
- 2005 testimony to the UK Parliament House of Lords Economic Affairs Committee;
- a 2006 interview on National Public Radio;
- a 2007 debate on National Public Radio;
- a paper published in 2007 in a climate contrarian-friendly social science journal, *Energy & Environment;*
- an article in 2008;
- an article in 2009; and
- an article in 2011.

The argument is based on several simple misrepresentations that a scientist of Lindzen's caliber should know are incorrect. For one, Lindzen assumes that as soon as humans emit greenhouse gases, the climate will immediately respond with 100 percent of the associated global warming. However, as already discussed, that's not how the global climate works. It takes many decades for the climate to fully respond to the energy imbalance created by the increased greenhouse effect due to the thermal inertia of the oceans. This is a basic concept in climate science that all climate experts understand, all climate experts except Richard Lindzen, perhaps.

Lindzen's misrepresentation is also based on only accounting for the influences that warm the climate, while ignoring those that cool the climate. For example, he accounts for all of the warming we expect from human greenhouse gas emissions, including methane and other greenhouse gases besides carbon dioxide, but entirely neglects the cooling effects caused by human sulfate aerosol emissions. Again, every climate expert knows that predictions made by climate models account for both warming and cooling effects, all climate experts except Richard Lindzen, perhaps.

By combining these two glaring errors, Lindzen has been wrongly claiming for about 25 years that climate models greatly overestimate global warming. Climate scientists and communicators, including myself, have pointed out Lindzen's obvious errors many times, and yet he continues to repeat this myth again and again nearly every year. And various media outlets continue to print his misrepresentations and to consider him a credible climate expert despite the fact that he makes these mistakes so basic that no undergraduate climate student should make. It's very difficult to imagine that Lindzen doesn't realize this argument is wrong, and yet he just keeps repeating it.

Lindzen has also argued over the past two decades that the climate sensitivity to the increased greenhouse effect is dramatically lower than that estimated by mainstream climate science research. He makes this argument in part through his incorrect claim that the planet has warmed less than expected; less warming would indicate lower sensitivity to the increased greenhouse effect, but of course this argument is false.

Lindzen also published a paper in 2009, claiming, based on satellite data, that climate sensitivity is exceptionally low; about a factor of 6 lower than the IPCC best estimate.[41] However, other climate scientists immediately identified several fundamental flaws in Lindzen's analysis. For example, a team led by eminent climate scientist Kevin

Trenberth found that Lindzen's low climate sensitivity result was heavily dependent on his choice of start and end points in the periods analyzed and that choosing different end points could yield completely opposite results with high climate sensitivity.[42] Lindzen's result was essentially based on cherry-picking convenient start and end points.

Two other papers published soon thereafter noted that Lindzen's result was based on examining data only from the tropics, whereas global climate sensitivity must be calculated using data across the entire planet.[43,44] Finally, a 2011 paper by Andrew Dessler concluded that Lindzen had assumed that changes in cloud cover were causing changes in temperature, whereas the evidence indicates the opposite is true, and ocean changes are causing changes in cloud cover.[45]

In 2011, Lindzen tried to publish a follow-up paper attempting to address the many valid criticisms of his 2009 paper. However, the follow-up paper failed to address most of the fundamental flaws identified by subsequent critical research. Lindzen submitted his response to the respected climate publication, the *Journal of Geophysical Research* (*JGR*). However, *JGR* declined to publish Lindzen's paper. He then submitted it to the *Proceedings of the National Academy of Science* (*PNAS*), which allowed him to select friendly reviewers, including physicist and outspoken climate contrarian William Happer.

However, *PNAS* noted that Happer is not a climate expert, and the other reviewer selected by Lindzen had formerly collaborated with him on similar research. Hence, they required additional reviews by unbiased, expert scientists. Those reviewers noted that among other problems with the paper, it failed to explain Trenberth's criticism that high climate sensitivity estimates could be obtained simply by choosing different starting and ending points for the analysis.[46] In fact, all four reviewers agreed that the paper was not of suitable quality for publication in *PNAS* and that its conclusions were not justified. As a result, like *JGR*, *PNAS* declined to publish the paper.

Lindzen was eventually able to get his paper published in an obscure Korean journal that likely had less stringent peer-review standards.[47] Submitting their papers to obscure and/or off-topic journals is one of the common ways climate contrarians publish shoddy scientific research, and it's an immediate red flag that their papers are fundamentally flawed. Editors of non-climate journals are less likely to be able to identify qualified climate experts to peer-review submitted papers, and hence, fundamentally flawed arguments are more likely to slip through their referee process.

Despite this history of consistently inaccurate and misleading climate arguments, Lindzen and his fellow climate contrarians continue to argue that global warming is nothing to worry about, and people continue to listen to them. Republican politicians have invited Lindzen to testify before U.S. Congress several times, and he's one of the most frequently cited scientists on climate contrarian blogs and conservative media outlets.[48] There's simply a double standard where contrarian climate scientists don't lose credibility, no matter how consistently wrong they're proven to be.

In fact, Lindzen was prominent in the press in January 2014, despite the fact that he retired from MIT in spring of 2013 to join a political advocacy group. Perhaps retirement has given Lindzen more time to chat with journalists. In any case, *The Weekly Standard* ran a long puff piece about Lindzen that, among other things, described him as "grandfatherly."[49] The entire piece focused on how Lindzen believes climate change is nothing to worry about, that changes in water vapor will dampen global warming, espousing the conspiracy theory that mainstream climate scientists are fear mongering in order to keep the spigot of federal scientific grant money open. It was a sorry piece of faux journalism that never even mentioned Lindzen's history of getting all of his climate-related arguments wrong. Fortunately, at the end of the long article, *The Weekly Standard* did include this quote from climate scientist Andrew Dessler:

> Over the past 25 years, Dr. Lindzen has published several theories about climate, all of which suggest that the climate will not warm much in response to increases in atmospheric CO_2. These theories have been tested by the scientific community and found to be completely without merit. Lindzen knows this, of course, and no longer makes any effort to engage with the scientific community about his theories (e.g., he does not present his work at scientific conferences). It seems his main audience today is Fox News and the editorial board of the Wall Street Journal.

Dessler nailed it. Nevertheless, the rest of the piece downplayed the risks associated with global warming based on nothing more than Richard Lindzen's opinion. Just a few days later, the television network CBS in Boston interviewed Lindzen about a new state of Massachusetts' plan to prepare for the impacts of climate change.[50] Predictably, Lindzen downplayed the risks associated with global warming, and CBS didn't investigate or mention Lindzen's history of being wrong on climate change. Not only is Lindzen a favorite of the conservative media, but he's also used by journalists to "balance" their articles with a perspective from "the other side." Unfortunately, that's just not how science works. It's not a democracy—arguments need to be supported by solid scientific evidence or they're cast aside. Giving Lindzen's opinion so much undeserved weight and prominence actually does media viewers and readers a disservice. This practice is known as "false balance."

The good news is that some media outlets are now beginning to hold Lindzen accountable for his history of being wrong. In February 2014, the NBC Sunday news show *Meet The Press* hosted a climate "debate" between Bill Nye the Science Guy and Tennessee Republican representative Marsha Blackburn. Blackburn tried to deny the expert scientific consensus on human-caused global warming by citing two contrarians, Judith Curry and Richard Lindzen. Fortunately, host David Gregory corrected Blackburn on the issue by pointing out that there is indeed an expert scientific consensus. However, it was disappointing that in one of the extremely rare instances that *Meet The Press* or any of the Sunday news shows discussed climate change, it was treated as a debate in which a politician was allowed to make a number of false statements about climate science and the expert consensus on human-caused global warming.

It's also unfortunate that it was posed as yet another of these one-on-one debates where fringe contrarian arguments are given equal weight to mainstream arguments. In May 2014, comedian John Oliver put together a great show on his HBO program *Last Week Tonight with John Oliver*. In the show, Oliver made several key points:

- Humanity's response to global warming has so far been a massive risk-management failure, or as Oliver wittily put it, "we've all proven that we cannot be trusted with the future tense."
- Public skepticism about global warming is irrelevant. As the great astrophysicist and science communicator Neil deGrasse Tyson says, "The good thing about science is that it's true whether or not you believe in it."
- The body of scientific evidence supports human-caused global warming.
- The media nevertheless continue to treat the subject as a debate, often with 1 person representing the majority consensus and 1 person representing the less than fringe minority.
- The debate should center on what to do about climate change; it's not about the science.

The pinnacle of the program was in the mock climate debate hosted by Oliver, initially involving Bill Nye the Science Guy versus a climate denier. However, in the interest of creating a statistically representative debate, Oliver brought out two more climate deniers and 96 more scientists. A video clip of the program went viral with over 4 million views, on top of the audience watching the program on HBO.

Soon after the Nye/Blackburn debate on *Meet The Press*, Ned Resnikoff at MSNBC ran an article quoting me in explaining Lindzen's history of being wrong.[51]

> "There is not consensus there," said Blackburn, who, as the current vice chair of the House Energy and Commerce Committee, has influence over domestic environmental policy. Blackburn cited two climate scientists to make her point: One who has been "wrong about nearly every major climate argument he's made over the past two decades," according to fellow environmental scientist Dana Nuccitelli, and another who recently said, "it's clear that adding more carbon dioxide to the atmosphere will warm the planet." The scientific consensus on climate change is accepted by 97% of climate scientists.

Resnikoff's piece was picked up and quoted by a few other media outlets. It's at least a positive development that more journalists are showing a willingness to look into the backgrounds of these few contrarian climate scientists who are constantly quoted and referenced by the conservative media in their efforts to deny human-caused global warming and oppose taking action to mitigate climate change. We need more journalists to hold climate contrarians like Richard Lindzen accountable for their history of constantly being wrong on climate science issues.

Those who continue to rely on Richard Lindzen to dispute the dangers of human-caused global warming are failing to learn from his history of being wrong. Again, George Santayana's famous saying bears repeating here:

> Those who cannot remember the past are condemned to repeat it.

4

The Formation and Growth of the Human-Caused Global Warming Consensus

By the late 1980s and early 1990s, a consensus was forming among climate science experts and in the peer-reviewed scientific literature that humans are the driving force behind the global warming we've experienced since the mid-20th century. This growing consensus has been examined by several different studies, starting with one written by science historian Naomi Oreskes of the University of California at San Diego (she has since moved to Harvard University), published in the prestigious journal *Science* in 2004.[1]

Oreskes reviewed the abstracts (paragraph-long summaries at the beginning of scientific papers) of 928 studies published in peer-reviewed scientific journals between 1993 and 2003 with the keywords "global climate change." She divided the papers into six categories: explicit endorsement of the consensus position (that humans are causing global warming), evaluation of climate impacts, climate mitigation proposals, methods, paleoclimate analysis, and rejection of the consensus position. Oreskes assumed that a paper evaluating climate impacts or discussing how to mitigate global warming implicitly endorsed human-caused global warming.

Of the 928 abstracts she reviewed, 75 percent fell in the first three categories, either explicitly or implicitly endorsing the consensus that humans are causing global warming. The other 25 percent fell in the paleoclimate (the study of past climate changes) and methods (e.g., researching how to improve measurements of climate change) categories. Oreskes did not find a single paper rejecting human-caused global warming in her sample of the peer-reviewed climate science literature.

A few months after Oreskes's study was published, a climate contrarian named Benny Peiser attempted to disprove her results. Peiser has since become the director of the UK political advocacy group Global Warming Policy Foundation (GWPF), which formed with the purpose of combating what their group describes as "extremely damaging and harmful policies" designed to mitigate climate change. Basically, they release shoddy scientific reports and use them to justify opposition to proposed policies to reduce greenhouse gas emissions.

Peiser repeated Oreskes's survey and initially claimed to have found 34 peer-reviewed studies rejecting the consensus on human-caused global warming (a less than 4 percent rejection rate). However, an inspection of each of these 34 studies by climate blogger Tim Lambert revealed that most of them don't reject the consensus at all.[2] Those few articles in Peiser's list that do qualify as rejecting the consensus were editorials or letters, not peer-reviewed studies. A group called Media Watch subsequently asked Peiser about those 34 studies, and Peiser admitted:[3]

> Only [a] few abstracts explicitly reject or doubt the AGW (anthropogenic global warming) consensus which is why I have publicly withdrawn this point of my critique.

When Media Watch pressed Peiser further, he admitted that only one article in his survey qualified as rejecting the consensus on human-caused global warming, and that was a committee report published in the Bulletin of the American Association of Petroleum Geologists (not peer-reviewed science). Oreskes's results were vindicated, confirming that the expert scientific consensus on human-caused global warming had formed by the mid-1990s, if not earlier, and confirming that peer-reviewed papers rejecting this expert consensus are very rare.

In 2009, Peter Doran and Maggie Zimmerman from the University of Illinois, Chicago, took a different approach to measuring the expert consensus. Rather than examine the peer-reviewed scientific literature, they conducted a survey of 3,146 Earth scientists and asked the question, "Do you think human activity is a significant contributing factor in changing mean global temperatures?"[4]

Overall, 82 percent of the Earth scientists answered "yes." Comparing the participants' responses to the level of expertise in climate science yields some particularly interesting results. Of scientists who were non-climatologists and didn't publish climate science research, just 77 percent answered "yes." In contrast, 97.5 percent of the scientists who actively publish research on climate change responded "yes." This suggests that as the level of active research and expertise in climate science increases, so does agreement that humans are significantly changing global temperatures.

This result was subsequently confirmed by a paper published in the *Proceedings of the National Academy of Science* in 2010, led by William Anderegg from Stanford University.[5] In this study, the authors compiled a database of 1,372 climate researchers who had either authored climate science assessment reports or signed statements about human-caused global warming. They then tallied the number of climate-relevant publications authored or coauthored by each of these researchers to evaluate their expertise in climate science. They set the criterion that in order to be considered a climate expert, a scientist was required to have published at least 20 climate-relevant publications, which trimmed their list to 908 researchers.

They found that the group of climate researchers who were "unconvinced by the evidence" of human-caused global warming comprised just 2 percent to 3 percent of the top 50 to 200 climate experts, in terms of having the most climate-related publications. The average climate researcher in the "unconvinced" category had published 60 climate papers, while the average researcher in the "convinced" category had published nearly twice as many, at 119. Approximately 80 percent of the "unconvinced" group was comprised of researchers with fewer than 20 climate-related scientific publications. Consistent with the findings of Doran and Zimmerman, the Anderegg group results show that greater expertise in climate science makes it more likely that a researcher will conclude that humans are causing global warming.

Despite this consistent picture with study after study showing a strong expert consensus on human-caused global warming, the American public has remained largely unaware of this fact. For example, a Pew Research poll in 2012 found that 57 percent of the U.S. public either disagreed or were unaware that scientists agree that the Earth is very likely warming due to human activity.[6] When asked what percentage of climate experts agree that humans are causing global warming, research by my colleague John Cook at the University of Queensland finds that the average American answer is 55 percent; far lower than the actual 97 percent. Cook has referred to this discrepancy as "the consensus gap."

In addition to the aforementioned studies examining the expert consensus on human-caused global warming, the scientific consensus is represented by various professional scientific associations. Dozens of scientific organizations endorse the consensus position that most of the global warming in recent decades can be attributed to human activities. The Academies of Science (organizations of the top scientists in each country) from 80 different nations all endorse the consensus on human-caused global warming. None reject it. Not a single professional scientific organization or National Academy of Science disagrees with the fact that humans are causing significant global warming.

And yet, the American public continues to believe that scientists are divided on this question. A 2014 survey conducted by scientists at Yale and George Mason Universities found that only 12 percent of Americans know that over 90 percent of climate scientists have concluded human-caused global warming is happening.[7]

How is that possible? While climate scientists and communicators have been struggling to make the public aware of this fact, climate contrarians and fossil fuel interest groups have expended considerable effort and resources to obscure the expert consensus. For example, as far back as 1991, Western Fuels Association conducted a $510,000 campaign whose primary goal was to "reposition global warming as theory (not fact)."[8] Republican political strategist Frank Luntz also emphasized the importance of this effort in an infamous memo that was leaked in 2002. In the section "Winning the Global Warming Debate—An Overview," Luntz emphasized the importance of obscuring public awareness of the climate science consensus in his first key strategy:[9]

> 1. ***The scientific debate remains open.*** Voters believe that there is ***no consensus*** about global warming within the scientific community. Should the public come to believe that the scientific issues are settled, their views about global warming will change accordingly. Therefore, ***you need to continue to make the lack of scientific certainty a primary issue in the debate***, and defer to scientists and other experts in the field.

Luntz was right. Social science research has shown that when people are aware of the expert consensus on human-caused global warming, they are more likely to support policies that will help solve the problem.[10] People who are aware of the expert consensus are also more likely to accept that human-caused climate change is happening.[11] These conclusions make sense—we can't be experts at every subject, so people tend to defer to the opinions of experts. If you feel ill, you visit a doctor. If your car isn't running well, you take it to a mechanic. If you want to know what's happening with climate change, you should listen to climate scientists.

When it comes to climate science, climate scientists are the experts. Polls have consistently shown that when people are asked who they trust as a source of information for global warming, climate scientists are at the top of the list. A 2012 poll conducted by the team at Yale and George Mason Universities found that 74 percent of Americans trust climate scientists when it comes to global warming.[12]

So the climate experts agree that humans are causing global warming, the public trusts climate experts on this subject, and people are more likely to support climate action if they are aware of the expert consensus. For those who oppose taking action to reduce greenhouse gas emissions from fossil fuel consumption, the math is clear. They have to convince the public that the expert scientific consensus doesn't exist, and so far they've done a very effective job of pulling the wool over our collective eyes. The climate denial campaign has mirrored the strategy and success of the tobacco industry, which managed to convince the public for decades that the science wasn't settled on the detrimental health impacts of smoking. They would find their own "experts" to publish biased research, claiming the health effects of smoking were unclear, and successfully convinced many people that scientific experts remained divided on the subject.

The media have played a key role in achieving this same goal of promoting "unsettled science" with regard to global warming. Consider Fox News' slogan: "Fair and Balanced." The network was built on the premise that the rest of the mainstream media has a politically liberal bias and that Fox News would "balance" it by providing a conservative perspective on the issues. Studies have shown that Fox News and other conservative media outlets like *The Wall Street Journal* heavily overrepresent the views of climate contrarians. In one sample, Fox News represented human-caused climate change doubters in 69 percent of stories.[13]

A Union of Concerned Scientists analysis of 2013 climate coverage by the three major American cable news networks similarly found that only 28 percent of Fox News' stories were factually accurate,[14] although that was an improvement from its 2012 climate reporting. In comparison, climate coverage on CNN in 2013 was 70 percent accurate, and MSNBC was 92 percent accurate. CNN's coverage was problematic in terms of the "consensus gap" as well, because it often hosted one-on-one climate "debates," creating the false perception that experts are evenly divided on the issue. The misperception has been worse yet for Fox News viewers, who are treated to entire programs disparaging climate science and advocating conspiracy theories that the scientific experts are all part of some sort of massive hoax.

The "false balance" approach, giving the views of less than 3 percent fringe minority of climate contrarians disproportionate media coverage, has seeped into other media outlets as well. Fearing accusations of bias, many newspapers and television news outlets will "balance" a mainstream scientific view by also interviewing a climate contrarian to present "the other side." The problem is that by disproportionately representing the small minority of climate contrarians, a false impression is created among the public that their numbers are greater than they are in reality, and hence the consensus gap is born.

The BBC provides a good example of this flawed false balance approach to climate reporting. In 2010–2011, the BBC Trust launched a review of the impartiality and accuracy of BBC science coverage, including an independent review by Steve Jones, Emeritus Professor of Genetics at University College London. Among his recommendations, Jones suggested that the BBC should stop giving "undue attention to marginal opinion" on climate science.[15]

However, in 2013, David Jordan, head of BBC editorial standards, told British members of Parliament that the network had decided not to follow Jones's key recommendations on climate change. In fact, BBC coverage of climate contrarians grew substantially in 2013. Most of the climate contrarians interviewed by the BBC were not climate scientists and had no particular expertise in the subjects being discussed. In a blatant display of false balance, the BBC chose these interviewees solely for their vocal contrarian opinions.

This may surprise Americans, who generally perceive the BBC as an accurate, unbiased, public news outlet. I know it surprised me to see such blatant disregard for actual balance in favor of false balance at the BBC, which I view as journalistic malpractice. Unfortunately, this is the reality in one of the world's most respected journalistic organizations. The BBC has interviewed bloggers and the head of the GWPF political advocacy group to present "the other side" in climate debates, despite the fact that these people have no climate science expertise whatsoever. When it comes to climate science coverage, the BBC is simply no longer a reliable journalistic source.

THE CONSENSUS PROJECT

In early 2012, a group of volunteers at Skeptical Science decided to update the climate literature survey conducted by Naomi Oreskes in 2004. Nearly a decade had passed, and a number of climate contrarians in the media were frequently claiming that the expert consensus on human-caused global warming was "crumbling." There was no evidence to support that assertion, so we felt it was a good time to test it with a new study of the peer-reviewed climate science literature, particularly given the general public perception that climate experts are divided on the causes of global warming.

My colleague John Cook, the founder of Skeptical Science, set up a system that would display abstracts of peer-reviewed climate science papers in random order for our volunteer citizen scientist participants to rate based on their position on the cause of global warming. Like Oreskes, we included papers using the key words "global climate change," but we also added "global warming" as a second search term. We expanded the search to include two decades' worth of peer-reviewed climate research, from 1991 through 2011.

Our team was comprised of about a dozen volunteers from all around the world. We had participants from the United States, Canada, Australia, New Zealand, Finland, Italy, and Germany, all of whom donated hundreds of hours of their free time to help complete this study.

Thanks to this large team of volunteers, we were able to expand our survey to a much larger set of studies than in Orsekes's 2004 paper. Adding in a second decade's worth of research and including papers with the keywords "global warming" ballooned the sample size from 928 in Oreskes's study to about 12,000 in our literature search. As it turns out, with concern about climate change and its impacts growing in recent years, the amount of research on the subject has also grown exponentially.

First we had to decide exactly how to conduct the study. We wanted to evaluate papers' positions on the cause of global warming, but internally we debated whether that would be enough. For example, should we use the same categories as Orekes's, or expand them? Should we just evaluate whether the surveyed studies agreed that humans are contributing to global warming, or should we include the specific percentage of global warming each study attributed to humans?

Surveys have shown that Americans believe scientists are divided on the simple question as to whether the planet is warming and whether humans are causing it. Thus, we decided our study should mainly focus on the simple question whether a study attributed global warming to humans without minimizing the human influence. However, I also wanted to keep track of studies that quantify the human contribution to climate change.

Our team of volunteers also gave us the resources to read a large volume of abstracts. While Oreskes assumed that a paper discussing climate impacts implicitly endorsed human-caused global warming, for example, we decided to carefully read each abstract and evaluate its position on human-caused global warming based solely on the language in the text. We also decided to make sure each abstract was rated by two independent participants to reduce human error. Where those two raters disagreed on a given abstract, they were allowed to revisit it to decide if they wanted to change their decision, or stand behind their original rating. Where the raters couldn't resolve their disagreement this way, a third rater acted as the tiebreaker.

In order to include information about the amount of global warming being attributed to humans, I proposed that we use the following categories for level of endorsement.

1. Explicit endorsement of human-caused global warming with quantification of the human contribution as responsible for greater than 50 percent of the global warming over the past 50 years.
2. Explicit endorsement of human-caused global warming without quantification.
3. Implicit endorsement of human-caused global warming. For example, papers that discuss mitigating global warming by reducing human greenhouse gas emissions, or papers that discuss greenhouse gases causing global warming without explicitly linking it to human activities.
4. No position on the causes of global warming. This turned out to be the most common category, because most papers related to global warming and global climate change don't bother to discuss the causes of global warming in their abstracts.
5. Implicit minimization or rejection of human-caused global warming. In this category we decided to include papers that acknowledge humans are causing global warming, but argue that humans are playing a small role. These papers imply that humans have caused less than half of the global warming over the past 50 years.
6. Explicit minimization or rejection of human-caused global warming without quantification.
7. Explicit minimization or rejection of human-caused global warming with quantification of the human contribution as responsible for less than 50 percent of the global warming over the past 50 years.

With these categories, we could have our cake and eat it too. They allowed us to address the simple question as to what percentage of peer-reviewed climate research attributed global warming to humans. At the same time, we could keep track of the papers that specified whether humans are responsible for more or less than half of the global warming that's happened over the past half century. Unfortunately, the nuances captured in these categories have been overlooked by many climate contrarians who have misrepresented our results; more on that will be discussed later.

We also categorized the abstracts in terms of the type of research the authors were doing, using similar categories to those in Oreskes's 2004 study.

1. Methods: papers that research technical aspects of climate measurement and/or modeling.
2. Mitigation: papers that examine how to slow global warming.
3. Paleoclimate: research into climate change from the past, prior to 1750 and often thousands or millions of years ago.
4. Impacts of climate change.
5. Not related to climate; accidentally included in our literature search.

While our team eventually agreed to proceed with the study using these categories, there was some disagreement about how some of the categories were defined. For example, I felt that since we know unequivocally that humans are responsible for the increase in carbon dioxide in the atmosphere (through various methods, including examining isotope ratios and by simple accounting, since we've actually emitted more carbon dioxide than has accumulated in the atmosphere), any paper discussing global warming caused by rising greenhouse gases should be considered an explicit endorsement of human-caused global warming. My colleagues disagreed, feeling that an explicit endorsement should include explicit language specifying that the global warming and/or increase in atmospheric greenhouse gases is human-caused. My colleagues outvoted me on that one, which made our results a bit overly conservative in my opinion. In fact we made several conscious decisions to ensure that our methods and results were conservative.

We were also conservative in deciding between the implicit endorsement and no position categories. We often came upon abstracts that seemed to clearly accept that humans are causing global warming, but that didn't use sufficiently specific language to fall into our implicit endorsement category. One of our volunteers actually decided to quit the project a few days after we began rating papers, because he felt uncomfortable placing papers into the "no position" category that he believed were in agreement with the consensus on human-caused global warming. However, the consensus among our team was that we should take a conservative approach in our ratings in order to shield our results from the inevitable criticism that would come from climate contrarians when they were published.

In 2012, Naomi Oreskes coauthored a paper titled "Climate Change Prediction: Erring on the Side of Least Drama?"[16] As the title suggests, the paper theorized that climate scientists tend to err on the side of taking a more conservative approach for several reasons, including to avoid criticism from climate contrarians. We took a similar "least drama" approach, choosing to be overly conservative in our paper categorizations in order to minimize opportunities for criticism. It's important to remember that an underestimate is just as wrong as an overestimate, even if the underestimate is less likely to be criticized. Nevertheless, we knew that our paper would come under heavy scrutiny because of the importance of the expert consensus on human-caused global warming, so we erred on the side of conservatism and least drama to ensure that our results would hold up to the most intense scrutiny.

And so we proceeded with our survey. John Cook set up a very well-designed system whereby a set of 10 randomly chosen climate papers from our literature search would display on the page. Our team of volunteers would mouse over the title of each, and the abstract would pop up. We would read each title and abstract and, based on the categories we had agreed upon, decide what category the paper belonged in for endorsement or minimization/rejection of human-caused global warming, and type of research.

It was a daunting task for a team of about a dozen volunteers. We had a set of about 12,000 abstracts to go through, rating each one at least twice. That meant we had to average about 2,000 abstract readings and categorizations each. This was all done in our spare time, on top of continuing to run the Skeptical Science website and blog and trying to carry on with the rest of our lives.

Personally, in the evenings when I arrived home after work, I would sit down for a few hours reading and categorizing abstracts. I was able to churn through 50 to 100 per night, a few nights per week, and some on the weekends as well. John Cook also cleverly set up a leaderboard to show who had categorized the most papers. This created a bit of friendly competition and motivated us to keep plowing through the abstracts.

After a few months I'd categorized nearly 1,800 abstracts, and our team had completed the initial categorizations of all the papers. We then began the process of addressing disagreements where the two individual ratings of a given abstract didn't match. If the two categorizations disagreed, we let both raters review the abstract a second time and either change or defend their initial categorization. If they still disagreed, we let a third rater read the abstract and cast the tiebreaker rating.

We were also concerned that once published, our results would be rejected and attacked. Climate contrarians have tried very hard to misinform the public about the expert consensus on climate change, and presumably we would arrive at a similar result as Oreskes, Doran, and Zimmerman and Anderegg et al. in finding a robust consensus. Although Skeptical Science is well respected by climate scientists for the scientific accuracy of its content, climate contrarians often try to label it as a biased advocacy site. In their eyes, any website that communicates the scientific evidence indicating that human-caused climate change is a major threat to the well-being of human society and civilization is a biased advocacy site.

While that characterization of Skeptical Science is completely inaccurate, climate contrarians also have significant influence in the conservative media like Fox News and *The Wall Street Journal*. We were a bit worried that our results would be rejected and smeared because the ratings were done by Skeptical Science contributors.

On top of that, we wanted to find a way to independently validate our results anyway. It's entirely possible that we could have subconsciously been biased in our categorizations of the scientific abstracts, as we ultimately pointed out in our paper. Scientific research is far more convincing when it's replicated by other independent methods.

We decided that in addition to taking an approach like that in Orekes's 2004 paper with the abstract ratings, we could test our results by using a similar approach to Doran and Zimmerman—ask the scientific authors to categorize their own papers. This approach is actually a hybrid of those previous two studies, because we weren't asking the scientists' opinions; we were asking them what their peer-reviewed research papers said about the causes of global warming.

This approach had several clear benefits. Who knows what the research says better than the scientists who wrote it? This would give us an unbiased second opinion to compare with our abstract rating results. On top of that, while our categorizations were based only on the paragraph-long abstracts, the authors could categorize their research based on the content of the entire papers. The results wouldn't be exactly comparable to our abstract ratings, because the scientists would be considering much more information than we were, but it would also allow us to determine what information we were losing by only considering the abstracts of the papers, and it would give us a second and independent estimate of the expert consensus on human-caused global warming.

The challenge was in contacting all of the authors of the 12,000 papers included in our survey. Most papers list a contact e-mail address for one or more authors, but we had to go through all the papers and extract those e-mail addresses by hand. This was mostly achieved through an incredible effort by one of our German volunteers, Baerbel Winkler. She plugged away at it, collecting e-mails from scientific papers for several months. Eventually Baerbel had added over 2,500 e-mails to our list, and with the rest of the Skeptical Science volunteers, we were able to put together a list of 8,536 e-mail addresses from scientists across 91 different countries.

John Cook put together another website listing the same categories as we used in our abstract ratings. He sent out an e-mail to the 8,536 scientific authors of the papers in our survey and asked them to categorize their own papers. About 1,200 scientists responded; many had multiple papers captured by our survey, so they categorized a total of over 2,100 papers in terms of their positions on human-caused global warming and the type of research they were conducting.

Once we finished the abstract ratings and received all the author self-ratings, we began to examine the data, and we found a remarkable result. In the abstract ratings, among the papers taking a position on the cause of global warming, 97.1 percent agreed with the consensus that humans are causing global warming. In the author self-ratings, 97.2 percent of papers taking a position on the issue agreed that humans are causing global warming. Although these were two entirely distinct and independent surveys, they arrived at almost identical results. They were also consistent with the previous results found by Oreskes, Doran, and Zimmerman, and Anderegg et al.

The biggest difference between our volunteer abstract ratings and the scientist author self-ratings was in the percentage of papers classified as "no position" on the cause of global warming. Not surprisingly, because the scientists were able to consider all of the information in their full papers, a much higher percentage took a position on the cause of global warming.

Scientific paper abstracts are very valuable real estate. Most journals have strict limits of just a few hundred words for their abstracts, and often they're the only information about a paper that's not hidden behind a paywall. Scientists have to capture all of the most important information from their papers in those brief abstracts. Few papers specifically investigate the causes of global warming, and the fact that humans are causing global warming has been well established in the scientific literature. Thus, most papers don't waste valuable words in the abstract stating something as obvious as "humans are causing global warming." It would be like a biologist saying in an abstract "species evolve through natural selection." It simply doesn't need to be said, because everybody reading the abstract knows it's true.

Thus, only about one-third of the 12,000 abstracts we read and categorized took a position on the cause of global warming. The other two-thirds fell into our "no position" category. However, in the scientist author self-ratings, about two-thirds took a position on the cause of global warming, while the other one-third fell into the "no position" category.

All of the Skeptical Science contributors to this project then had to decide how to present this information and data in a scientific paper. Should we focus on all the results, including the "no position" category, should we focus on the papers taking a position on the cause of global warming, or should we focus on the papers taking a position that quantified the human contribution to global warming? There was much debate on this and other questions. Ironically, we had a very difficult time reaching a consensus about how to write our consensus paper.

In the end, while we included all the data and categories in our paper, we decided on the second option, focusing our discussion on the papers taking a position on the cause of global warming. We wanted to know the answer to that question—what does the peer-reviewed climate science research say about the cause of global warming? This was the simple question that the general American public was getting badly wrong.

It also turned out that there were relatively few abstracts and papers that quantified the human contribution to global warming. That specific question simply isn't relevant to most climate research. For example, if you want to research the effects of climate change on mountain goats in the Himalayas, what's causing that climate change doesn't matter to your research. Scientists tend to be very specialized in their research, and a scientist researching climate impacts on mountain goats would leave it up to other specialists to determine what's causing that climate change.

So in answering the question "What percentage of peer-reviewed climate science papers agree that humans are causing global warming?" studies that don't answer the question (the "no position" category) because it's not necessary to their research aren't relevant. We wanted to know, of the papers that do answer the question, how many agree and how many disagree. Answering this question would also make our results comparable to the previous consensus studies by Oreskes, Doran, and Zimmerman and Anderegg et al.

We first looked at the results of the abstracts that we had categorized as taking a position on the cause of global warming, either attributing it to human activity or minimizing or rejecting the human influence on climate change. All together these categories gave us a nice sample size of over 3,900 abstracts with which to answer the consensus question. Of those abstracts, 3,896 endorsed human-caused global warming while just 78 rejected or minimized it.

However, there remained a question about the "no position" abstracts. Some of our team members argued that there could be a significant number of abstracts in this category arguing that we simply can't determine what's causing global warming. Those papers would be taking a position on the cause of global warming, calling it uncertain, and should be included in the consensus calculation. The problem was that we hadn't broken out the "no position" category in terms of abstracts taking no position on the cause of global warming, and those saying the causes are uncertain. None of us remembered very many abstracts saying the causes of global warming were uncertain (I couldn't remember any out of the nearly 1,800 abstracts I had categorized), but we didn't want to leave this issue unresolved.

We decided to take a random sample of 1,000 abstracts in the "no position" category to see how many would fall into two new subcategories, "no position" and "uncertain" about the causes of global warming. After a few more days of reading and rating these abstracts, we found just 5 of the 1,000 that could be interpreted as saying that the causes of global warming are uncertain. All of the rest of the "neutral" abstracts simply didn't take a position on what's causing global warming. Extrapolating this sample to the 7,930 in the "no opinion" category gave us an estimate of 40 "uncertain" abstracts.

Putting these 40 uncertain abstracts together with the 3,896 abstracts endorsing human-caused global warming and 78 rejecting or minimizing it gave us the overall consensus of 97.1 percent. We also looked at the abstracts and papers that quantified the human contribution to global warming. There were relatively few of these; just 75 of our abstract ratings and 237 of the scientific author self-ratings captured in our literature search specified how much of the observed global warming is being caused by humans. Relatively few studies investigate this specific question, and those who don't investigate it tend not to address the question in their abstracts or papers. Nevertheless, we found that among those who did quantify human-caused global warming, 87 percent of abstracts and 96 percent of author self-rated papers stated that humans are the primary cause of the current global warming. The abstracts were a small sample size, so the self-rated papers represent a more reliable result for the consensus among papers explicitly quantifying the human contribution to global warming.

The 96 percent consensus in this category is almost identical to the overall 97 percent consensus (including explicit endorsements that didn't quantify the human contribution, and implicit endorsements). It's also important to remember that any papers implicitly or explicitly minimizing the human contribution to global warming were put in the "rejection" categories, so both the 96 percent and 97 percent figures can be described as the expert consensus that humans are the main cause of global warming since the 1950s. The 96 percent is explicit, but a smaller sample size, while the 97 percent includes implicit endorsements of the consensus, but gives us a much larger sample size.

So our study clearly debunked the "no consensus" myth, but another related misconception was also gaining popularity. Before we began our survey, we had noticed climate contrarians claiming more and more frequently that the expert consensus on human-caused global warming was crumbling. These claims were usually made after the publication of a single contrarian paper or were based on the contrarians' biased misperceptions. There was no solid evidence to support the assertion, but it was another consensus-related claim that we could test in our study.

Remember, we had examined papers published over the previous two decades, between 1991 and 2011. Thus, our newly created database included all the necessary information to evaluate whether the expert climate consensus was crumbling, steady, or growing. After we finished looking at the overall consensus results, we broke the data out year by year to see how the consensus had changed over time (Figure 4.1).

Not surprisingly, the data shows no evidence whatsoever of a crumbling consensus. In fact, there was a slow but steady growth in the expert consensus on human-caused global warming between 1991 and 2011. What we didn't expect to see was that the consensus had already formed by the early 1990s. In our abstract ratings (which are steadier and less noisy than the scientist author self-ratings because the sample size is bigger), the consensus was already well over 90 percent in 1991. It grew slowly but steadily over the next two decades, reaching 98 percent in both the abstract and scientist self-ratings in 2011.

So where did the crumbling consensus claim come from? Like many climate myths, it was based on wishful thinking at best and intentional misinformation at worst. As Chip and Dan Heath discussed in their book *Made to Stick*, climate science myths are often effective and pervasive because they are "sticky"—simple, concrete, and seemingly

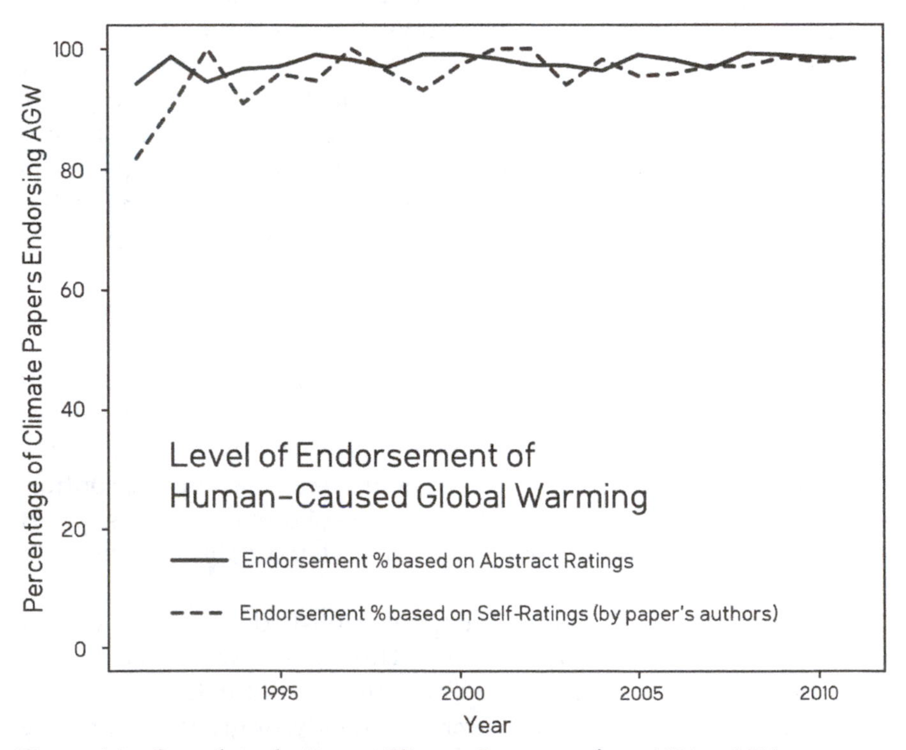

Figure 4.1 Growth in the Expert Climate Consensus from 1991 to 2011

credible.[17] The crumbling consensus myth checked all of those boxes. The challenge in debunking sticky myths is to replace them with even stickier facts. There are several potential backfire effects that can act to actually reinforce the myth in peoples' minds, as discussed by John Cook and Stephan Lewandowsky in their *Debunking Handbook*.[18] For example, going into too much detail and information when explaining why the myth is wrong can actually cause people to remember the myth more vividly and forget the facts debunking it. This is known as "the overkill backfire effect."

Fortunately, the Skeptical Science team had extensive experience in effectively debunking climate myths. That's what the site was built for, and it's a subject John Cook in particular has researched extensively. Perhaps the most important key to effective communication is KISS—keep it simple, stupid. This is especially true with the popularity of social media today. A simple message that can be easily communicated, for example, within the 140-character limit of a Twitter tweet, can potentially reach a large audience.

Our consensus study also lent itself to several simple but important messages. First, there is a 97 percent consensus in the peer-reviewed climate science literature that humans are causing global warming. Second, that consensus is growing stronger over time. We were also able to create several infographics to communicate the main results of our study. People love infographics because when done right, they can simply and effectively communicate useful information, and infographics can easily be shared via social media.

In building up credibility working on Skeptical Science over the years, we had also created a long list of contacts among journalists on other blogs and in the media. The "sticky" nature of our results, the importance of communicating the consensus, and our ability to get the word out by using our media contacts to communicate our simple message made for an excellent opportunity.

We spent a good six months to a year putting the paper together after we had completed the abstract ratings and received the author self-rating results, making sure it was as rock solid as possible with no significant weaknesses that climate contrarians could latch onto. We submitted the paper to a high-impact journal, *Environmental Research Letters*, which we chose because it was a well-respected journal that publishes a lot of climate science research and because it offered the option to publish the paper as open access. Most scientific journals will accept papers and publish them at no fee to the scientists, but accessing the full paper will cost readers somewhere in the ballpark of $30.

We wanted our paper to be free for everybody to access and read, and some journals offer the option for the authors to pay an up-front fee to make the paper freely available. We asked the readers of Skeptical Science to help us raise the funds, and they crowdsourced the open access fee in under 10 hours.

After we submitted the first draft of our paper to the journal, it was sent out to several climate expert referees for the peer-review process. A few months later we received their comments, and all were very positive. The reviewers made some good suggestions that resulted in relatively minor changes, but helped make the paper even stronger. Once we submitted a revised version of the paper incorporating those suggestions, the editor at *Environmental Research Letters* accepted it for publication. We raised and paid the fee to make it open access, and the wheels moved rapidly into motion.

We had also been contacted by a public relations company named SJI Associates that wanted to do some pro bono work for Skeptical Science because their staff appreciated the quality of the site, the importance of the climate issue, and the volunteer nature of our contributor team. The results of our soon-to-be published consensus paper provided the perfect opportunity to collaborate with SJI Associates to help maximize the impact and reach of our results. SJI Associates created a great website called TheConsensusProject.com, which explained some basic climate science with the aid of creative animations and also featured the results of our study along with some nice infographics.

We wanted to make our data and results as transparent and replicable as possible. We knew that climate contrarians would attack our study, claiming that the Skeptical Science volunteers categorizing the papers were biased. The scientist author self-ratings arriving at the same 97 percent consensus result gave us an ironclad response to that inevitable attack, but we also wanted to allow individuals the opportunity to check our results for themselves. That way our results would be unassailable, because anybody could verify them for themselves. John Cook set up a website where he imported all the climate paper abstracts that we had categorized during our survey. He set up the same system as we had used, displaying titles randomly and allowing people to read the abstracts by dragging the mouse over the title of the paper.

In May 2013 we had everything ready to go. Our paper was about to be published in a respected peer-reviewed journal. We had found a 97 percent consensus in the climate science literature using two independent methods, volunteer abstract ratings and scientist author self-ratings, and our paper would be free for anybody to download to see exactly what we did. We had a system with which anybody could check our results for themselves. We had the slick new website created for us pro bono by SJI Associates. And we had a very news-friendly story: a team of citizen science volunteers from around the world conducting the most comprehensive survey ever done of the peer-reviewed climate science literature, finding the same result using two independent methods. Most importantly, we had a simple, concrete, sticky message with the 97 percent expert consensus on human-caused global warming.

John Cook and I got to work, sending information about our paper and associated resources to all of the climate and environment journalists whose contacts we had built up over the previous several years. Because Skeptical Science was such a useful and trusted resource for fact-checking climate claims, we had previously been contacted by many journalists, and most other environmental journalists to whom we sent information about our study were familiar with Skeptical Science and its solid scientific credentials. All together, it made for an appealing story, and we made sure a lot of environmental journalists were aware of it and would have time to prepare stories before the paper was published by *Environmental Research Letters.*

On May 15th, 2013, the paper was published. Our media outreach turned out to be more successful than we ever dreamed. Newspapers from around the world ran stories about our study. John Cook and I each did several radio interviews. John was also interviewed on CNN, and I talked about our results in a panel on the Al Jazeera show *Inside Story.* The most exciting and probably most influential communication of our results came via social media. President Obama's Twitter account published two separate Tweets about our study. In fact, President Obama's Tweets launched a second wave of media stories about our paper.

About five months later, our paper became the most downloaded in all of the *Institute of Physics* scientific journals (*Environmental Research Letters* being one of the *Institute of Physics* journals), with over 100,000 downloads. It has now been downloaded over 270,000 times; more than double the second-most downloaded paper in all *Institute of Physics* scientific journals. The editors and publishers at *Environmental Research Letters* voted it their best paper of 2013. It was also the 11th most talked about academic paper of 2013 according to Altimetric, which published a review of the 100 academic papers "that received the most attention online and the conversations that happened around them." Not bad for a nerdy scientific paper written by a bunch of citizen scientist volunteers.

As we expected, despite all the lengths we went to in an effort to make our results as transparent and replicable as possible, climate contrarians mounted numerous attacks. One of the main attacks claimed that the 97 percent consensus was so weak, simply stating that humans are causing *some* global warming, that even the climate contrarians agree with it. For example, climate scientist and contrarian Roy Spencer of the University of Alabama at Huntsville claimed in testimony to U.S. Congress and later in media interviews and on blogs that he would be included in the 97 percent despite being a "skeptic."

There are several problems with Spencer's assertion. The biggest flaw is that five of Spencer's papers were included in our abstract ratings: four of those were categorized as "no position," and the fifth was categorized as implicitly minimizing or rejecting the human contribution to global warming. In other words, Spencer isn't in the 97 percent; his papers are in the less than 3 percent. His assertion to Congress was flat-out wrong.

Moreover, rather than make false claims to U.S. Congress and journalists, Spencer could have easily checked how we categorized his abstracts. All of our results were published online, and the interactive ratings system made them easy to check by doing keyword searchers, for example, by author name. It only took me about 30 seconds to see that Spencer's abstracts did not fall into 97 percent consensus categories.

Spencer and his fellow climate contrarians failed to grasp the nuance of our study. In order to fall into the less than 3 percent of contrarian papers, an abstract only had to implicitly minimize the human contribution to global warming. On top of that, we also looked at papers that quantified the human contribution to global warming. As previously noted, 87 percent of abstracts and 96 percent of scientist self-rated papers that quantified the human contribution to global warming stated that humans are the main cause of climate change over the past 50 years.

Thus, in addition to the 97 percent consensus that humans are causing significant global warming, our results included a 96 percent consensus that humans are the main cause of global warming over the past half century. The claims made by Spencer and his fellow contrarians that they would be included in the consensus were wrong on multiple levels, and Spencer could have fact-checked his claim prior to making it on congressional testimony with a simple 30-second search of our abstract ratings website database.

RICHARD TOL ACCIDENTALLY CONFIRMS THE 97 PERCENT CONSENSUS

The attacks on our paper didn't stop there, of course. From the outset, economist Richard Tol seemed determined to find something wrong with the methods we used in our abstract ratings. Strangely, Tol admitted that he believes the expert consensus on human-caused global warming was real and correct, but nevertheless spent the next year trying to poke holes in our methodology.

We never were able to figure out what Tol's motivations were in attacking our study. Perhaps he saw the attention our results had received and thought he could raise his own profile by taking us down. Perhaps he enjoyed the praise lavished on him by climate contrarians whenever he spoke critically about our paper. He explicitly admitted that the consensus is "of course in the high nineties,"[19] but decided that nit-picking whether 97 percent is exactly right was a good use of his time.

Richard Tol is a fairly well-known economist with an extensive record publishing papers on the costs of climate change, and so his credentials seemed to lend credibility to the attacks on our paper. In fact on the BBC show *Sunday Politics*, host Andrew Neil claimed in July 2013 that our results had "of course been substantially discredited" and that "Professor Richard Tol . . . has disassociated himself from that and said it's not reliable."[20] Of course, Tol was never associated with our study, so the claim that he had disassociated himself from it was entirely nonsensical.

All of these negative claims made on the BBC about our peer-reviewed study were based on nothing more than comments Tol had made on the Internet. The BBC was soon criticized for frequently giving airtime to this sort of factually inaccurate climate contrarianism, but as previously noted, the organization has unfortunately decided to pursue a strategy of false balance rather than hold itself to a standard of factual accuracy. It has been disappointing to see the once highly respected BBC's journalistic standards fall to a level similar to those of Fox News when reporting on climate change.

Our study seemed to be an obsession of Tol's, as he eventually submitted five papers critiquing our approach, four of which were rejected by three separate journals. Normally a critique of a paper should be published in the same journal that published the paper. However, *Environmental Research Letters* is a high-impact journal, rejecting 65 percent of papers it receives as submissions, so it has a high standard of quality. Tol's paper didn't come close to meeting that standard. It was full of unsupported, unsubstantiated, incorrect statements, and the journal editor immediately rejected his first submission, writing,

> It is in a large part an opinion piece. . . . I do not see that the submission has identified any clear errors in the Cook et al. paper that would call its conclusions into question—in fact he agrees that the consensus documented by Cook et al. exists.[21]

Tol revised his critique and submitted it to *Environmental Research Letters* again; the second time, the editor sent it out to climate expert referees for peer-review. Tol posted the reviews on his blog, and the reviewers were not kind, saying, for example,

> Rather than contribute to the discussion, the paper instead seems oriented at casting doubt on the Cook paper, which is not appropriate to a peer-reviewed venue, and has a number of important flaws. . . . Many of the claims in the abstract and conclusion are not supported by the author's analyses. . . . Language is overly polemical and not professional in some areas. At times in the introduction and conclusion, the language used is charged, combative, not appropriate of a peer-reviewed article and reads more like a blog post. This does not serve the paper well and reflects poorly on the author.[22]

Based on these critical comments, *Environmental Research Letters* once again rejected Richard Tol's comment. A few months later, another journal called *Science & Education* did publish a critique of our paper, which was based on blog comments made by Christopher Monckton, who is essentially a climate contrarian celebrity and nonscientist.[23] The critique was a response to a comment by Daniel Bedford and John Cook,[24] which the same journal had published just two days after submittal, so apparently this particular journal's peer-review process leaves much to be desired.

Indeed, the Monckton critique was exceptionally poor. It basically argued that only 65 abstracts out of the 12,000 in our survey quantified that humans have caused most of the global warming over the past 50 years, and 65 of 12,000 is a fraction of a percent; therefore, the consensus was more like 0.5 percent rather than 97 percent. However, this argument neglects the fact that our literature search using the keywords "global warming" and "global climate change" was quite broad, and many abstracts and papers simply didn't take a position on the cause of global warming. That doesn't mean they disagree that humans are causing global warming; it simply means their abstracts and/or papers are not relevant in answering that question.

Using this logic, you could argue that there's a less than 1 percent consensus on evolution, that the Earth revolves around the sun, or that the Earth isn't flat. Moreover, using Monckton's approach, only 0.08 percent of abstracts would qualify as minimizing or rejecting human-caused global warming. That's not an argument climate contrarians should want to make.

After the two rejections by *Environmental Research Letters*, Tol submitted his paper to three other journals, two of which he reported rejected it for being outside of their scope. On the fifth try, a journal called *Energy Policy* finally accepted Tol's critique of our paper. As the journal name suggests, a critique of our paper does not fall within the scope of *Energy Policy* either, which describes itself as[25]

> an international peer-reviewed journal addressing the policy implications of energy supply and use from their economic, social, planning and environmental aspects.

A paper critiquing the methodology of another paper that quantifies the consensus in the peer-reviewed climate science literature on human-caused global warming has essentially nothing to do with the policy implications of energy supply. In fact, in his paper Tol specifically stated:

> Consensus has no academic value (although the occasional stock take is valuable for teaching and guiding future research) and limited policy value.[26]

However, he also tacked on four sentences at the end of the paper that weren't present in previous drafts, talking about its "policy implications." This was a rather transparent attempt to make his paper fit within the scope of *Energy Policy*, and for some reason the journal editors bought it.

When we looked at his paper, the first thing we noticed was that Tol hadn't addressed any of the constructive criticism that the *Environmental Research Letters* reviewers had given him when that journal twice rejected his submission. They identified 24 problems or areas where the paper could be improved; Tol ignored every single one. His *Energy Policy* paper had all the same errors, and he added some big new ones. He had even referenced Monckton's terrible paper, as well as various climate contrarian blogs, including their discussions of material that had been stolen during a hack of the Skeptical Science private discussion forum. Tol took some comments by one of our volunteers, Andy Skuce, about feeling déjà vu because language in the abstracts of climate papers are sometimes similar, and claimed they supported his hypothesis that our volunteers suffered from "rater fatigue."

Aside from the referenced material being stolen and thus totally unsuitable for an academic paper, Tol had also badly misrepresented Andy's private comments, which had nothing to do with fatigue. In fact, our ratings were done at our own leisure without any deadlines. Fatigue was never a problem. Our team also became more adept at rating papers over time, because we became more experienced after reading hundreds and hundreds of abstracts. If anything, we got better at rating papers over time as we became more experienced, not worse. Even after Andy clarified his comments and complained that Tol was distorting them, Tol continued to stand behind his misinterpretation of those comments. His bias could not have been more transparent.

Tol's paper also argued that rather than reading and categorizing every abstract individually based on its precise wording, we should have taken all the abstracts dealing with climate impacts and assumed that they all had no position on the cause of global warming. This would be similar but opposite to the approach taken by Naomi Oreskes in her 2004 study, where she assumed that papers investigating the impacts of climate change were implicitly assuming that humans are causing global warming. Oreskes's approach makes more sense than Tol's suggestion; after all, why worry about the impacts of global warming if it's not human-caused and therefore probably won't continue?

However, our approach of reading every single abstract and categorizing them based on their specific language was much more thorough and precise than making sweeping general assumptions. Tol's criticism would have involved making our study less thorough and precise. It simply made no sense.

We also found it funny that in his critique, Tol explicitly stated:

> There is no doubt in my mind that the literature on climate change overwhelmingly supports the hypothesis that climate change is caused by humans. I have very little reason to doubt that the consensus is indeed correct.[27]

Tol was obsessed with publishing a critique of a paper whose conclusions he believed were true. In any case, his critique was also based on faulty statistics. He only quantified how his criticisms impacted our final result in one spot, claiming that errors in our ratings would change the consensus result from 97 percent to 91 percent. Now, 91 percent would still be an overwhelming expert consensus on human-caused global warming, but Tol came up with this "correction" to our conclusion by making an obvious mathematical error.

As previously discussed, our survey included a reconciliation process to address abstracts where the two independent raters disagreed on which category the paper belonged in. By looking at the statistics of the reconciliation process, it's possible to estimate the remaining error, because our raters were human after all and thus made some mistakes.

Tol estimated that the remaining error was 6.7 percent, which is probably an overestimate, but not totally unreasonable. However, he got lazy and sloppy in his statistical analysis. In our reconciliation process, 55 percent of the changed ratings were "toward greater rejection" and 45 percent "toward greater endorsement." Tol thus assumed that of the 6.7 percent of abstracts in the "no position" category that he believed were incorrectly rated, 55 percent of those should be rejections and 45 percent should be endorsements.

He didn't bother to check how the ratings changed for each category during the reconciliation process. Most of the changes "toward greater rejection" were from implicit endorsement to "no position" or from explicit to implicit endorsement. For the "no position" category, 98 percent of changes were to endorsement categories, and just 2 percent were toward rejections. That makes sense when you think about it, because less than 3 percent of all climate papers reject or minimize human-caused global warming. There's no reason to expect 55 percent of incorrectly rated "no position" papers to reject the consensus—in reality there just aren't that many rejection papers.

Since "no position" was the biggest category for our abstract ratings, and because Tol assumed that 55 percent of the incorrectly rated papers should move to rejections rather than the correct figure of 2 percent, he effectively conjured up 300 rejection papers, pulling them out of thin air with nothing more than a math error. It's easy to see this is wrong. In our full sample of 100 percent of the abstracts, we identified only 78 papers rejecting the consensus. In the mere 6.7 percent that Tol believed were in error, he nearly quadrupled that figure, conjuring up another 300 rejection papers.

My background is in physics, and in that field you learn quickly to always do a "sanity check" on your calculations. For example, if you're working on a problem and conclude that Mars has a mass 1,000 times larger than that of Earth, you think about how realistic that is, realize you must have made a mistake somewhere, and go back to check your math. Maybe economists don't do that because economics isn't a physical science with intuitive answers. For whatever reason, Tol and the referees who peer-reviewed his paper obviously didn't do a sanity check on his results, because they fail badly. You don't have to understand statistics to realize that if you've quadrupled the number of rejection papers with just a 6.7 percent sample, your math has gone awry somewhere.

Ultimately we also had the trump card of the scientist author self-ratings. You could throw our volunteer abstract ratings out and still be left with the 97 percent expert consensus that humans are causing global warming and the 96 percent expert consensus that humans have caused most of the global warming over the past 50 years, purely from the scientists categorizing their own papers. In fact, none of the criticisms of our paper have ever addressed the author self-ratings. They have all tried to find fault in our volunteer ratings of the papers' abstracts, overlooking the fact that we also found a 97 percent consensus in the peer-reviewed climate science literature using a second, completely independent method.

When we corrected for Tol's error, accounting for the ways in which the ratings had actually changed in each category during the reconciliation process, we found the consensus estimate was robust at 97 ± 1 percent. That was just the tip of the iceberg; in the end we identified 24 errors in Tol's paper. Many of these were the same errors identified by the *Environmental Research Letters* reviewers, some those reviewers just hadn't identified, and some were new to the *Energy Policy* version of his paper.

Unfortunately, our dealings with the *Energy Policy* editors were not very positive. They gave us the opportunity to publish a response to Tol's paper, but would only allow us 1,000 words and two weeks to respond. One thousand words were barely enough to scratch the surface of Tol's error-riddled paper. However, we used that opportunity to detail the math errors that caused Tol to underestimate the consensus at 91 percent and then referenced a document that we published on Skeptical Science detailing the rest of his errors, at sks.to/TolReply.

John Cook politely asked the *Energy Policy* editor a number of questions to make sure our submission met the journal's requirements. Since *Energy Policy* had given us only two weeks to respond, we wanted our submission to go smoothly. Cook also requested that the journal allow us more than 1,000 words, for example, in a supplement to our paper, but the editor refused. Eventually she told him to "stop harassing" her and just submit the document. We later found out that she was a consultant for the oil industry. Given that the editors had published Tol's error-riddled paper despite the subject clearly being outside the journal's scope, did not give us a fair opportunity to fully respond, and also allowed Tol the last word with a comment on our response, we were left with the feeling that the *Energy Policy* editor was not acting in good faith.

We were surprised when Tol's paper went largely ignored after it was published. Just before it was published, Republicans had invited Tol to testify before the U.S. House of Representatives Congressional Committee on Science, Space, and Technology about the latest Intergovernmental Panel on Climate Change report. The subject of the 97 percent expert consensus on human-caused global warming came up a few times, and Tol made several disparaging comments about our paper. Thus, we had every reason to expect that once his critique was actually published, the conservative media would go berserk about it. However, we got out in front, pointing out that Tol agreed the consensus was real, in the high nineties, and correct. In blog posts, we also detailed his 91 percent math error and the other 23 mistakes in his paper. These points may have discouraged the conservative media from covering the story. On the contrarian blogs that wrote about Tol's paper, many commenters expressed dismay that Tol admitted the consensus is real. For whatever reason, Tol's terrible paper thankfully got little attention.

THE SOCIAL SCIENCE CLIMATE COMMUNICATIONS DEBATE

The results of our 2013 consensus study also triggered a debate among social scientists as to how best convince the general public about the threat of human-caused global warming. Our results helped address the consensus gap, where the average American thinks only 55 percent of climate experts agree on human-caused global warming, far lower than the actual 97 percent. However, Yale Professor of Law and Psychology Dan Kahan has argued that the consensus is not an effective communications tool, at least by itself. Kahan believes that peoples' positions on global warming are dictated by their political and cultural biases, and for any piece of factual information to help change their opinions (including the existence of the expert consensus), you first have to find a way to break through or get past that cultural bias.

It's certainly true that research done by Kahan and other social scientists has clearly shown that cultural biases are one of the main factors determining peoples' perception of climate change, perhaps even the single biggest factor. In essence, liberals feel as though they're on Team "global warming is a problem caused by humans" while conservatives identify with Team "no it's not." Kahan feels that people will take any new information and pass it through their cultural filter; if it conforms to their cultural identity, they'll accept it. If not, they'll just reject it. In fact, Kahan argues that giving people information that doesn't conform to their cultural identity (like the 97 percent consensus) may just act to polarize them further.

However, research done by John Cook and Stephan Lewandowsky has shown that when people are made aware of the expert consensus on the subject, across the political spectrum they become more likely to accept human-caused global warming (with the exception of extremely conservative Americans). An interesting 2014 study by social scientists at Yale and George Mason Universities also tested various methods of communicating the 97 percent consensus to audiences of various political persuasions.[28] They tried conveying the consensus as a simple statement, as a metaphor (e.g., "If 97 percent of doctors concluded that your child is sick, would you believe them? 97 percent of climate scientists have concluded that human-caused climate change is happening"), and as an illustration in pie chart form. The scientists found that all of these methods increased the subjects' perception of the consensus, but the biggest change was among Republicans who were exposed to the pie chart. Their perception of the expert consensus increased by a whopping 27 percent. So conveying this expert consensus information, especially in pie chart form, seems to increase awareness even among political conservatives.

Kahan remains unconvinced by these data. He argues that the experiments aren't real-world results, and in the real world, climate realists have been communicating the existence of the expert consensus for nearly two decades. If the expert consensus is an effective message, then why hasn't it convinced the public that human-caused global warming is real and a problem?

I believe the answer to that question lies in the fact that climate contrarians have at the same time been working very hard to dispute the existence of the expert consensus over the past two decades. The media have been complicit in helping them achieve this goal through the practice of false balance. Climate contrarians make up less than 3 percent of climate experts, but they receive a much larger proportion of media coverage, especially in the conservative media like Fox News and *The Wall Street Journal*, but also in media outlets that should be unbiased and yet still give climate contrarians a disproportionate amount of coverage, like the BBC. When people turn on the television or read a newspaper and frequently see climate contrarians saying global warming isn't happening or isn't caused by humans or isn't a problem, it gives them the false perception that the experts are divided on these issues.

A perfect example of this behavior involved a 2013 study examining the opinions of the American Meteorological Society (AMS) about global warming.[29] It's long been known that meteorologists have a relatively low acceptance of human-caused global warming. In the survey of Earth scientists conducted by Doran and Zimmerman in 2009, meteorologists had the second-lowest acceptance at 64 percent, behind only economic geologists (47 percent), who have a clear conflict of interest, often working for fossil fuel companies. The 2013 AMS survey tried to determine what factors were causing their relatively high rejection of human-caused global warming.

The study found that the factor that best predicted meteorologists' positions on global warming was their awareness of the expert consensus. The second-best predictive factor was their political ideology. Coming in third was their expertise in climate science. While meteorology is a somewhat related field, only 13 percent of AMS members described climate as their field of expertise. Among those climate experts, 93 percent agreed that humans have contributed significantly to global warming over the past 150 years. Among nonexperts, the answer was much lower.

Several media outlets misrepresented this study in their reporting. For example, *Forbes* business magazine ran a piece by James Taylor of the conservative Heartland Institute think tank, who claimed the low acceptance of human-caused global warming among AMS members disproved the expert consensus. I interviewed Neal Stenhouse, lead author of the study, who was not pleased with the Heartland Institute's distortion of his results. Stenhouse told me:

> Mr. Taylor's claims are highly misleading, but we expect that from someone with a long history of distorting the truth about global warming. We found high levels of consensus on global warming among the climate experts in our sample. You only see low levels of consensus in the sample if you also look at the views of people who are not climate experts.[30]

Stenhouse's study again showed that perception of the expert consensus is a key factor in determining whether people accept human-caused global warming. We also know that lack of awareness of the expert consensus isn't purely due to political and cultural biases. Although the consensus gap is larger among American conservatives (who think the expert consensus is about 35 percent to 50 percent, according to John Cook's research), there's also a consensus gap among American liberals (who think the expert consensus is about 55 percent to 70 percent). That 20 percent difference between conservatives and liberals can be explained by political biases, but if that were the only explanation, then American liberals would respond that 97 percent of climate experts agree on human-caused global warming. Thus, a big chunk of the consensus gap isn't political; it's due to something else. I believe it's most likely a reflection of media false balance. The fact that only 12 percent of Americans are aware that the consensus is above 90 percent is further evidence that cultural and political biases aren't the only problem at play; liberals and moderates are unaware of the consensus too.

This poses a major challenge for climate communicators like myself and my Skeptical Science colleagues. We can repeat the existence of the consensus until we're blue in the face, but as long as media outlets continue to give disproportionate representation to climate contrarians, the message won't stick. And if people remain relatively unaware of the expert consensus, they're unlikely to grasp the urgency of the climate problem and thus are unlikely to demand that our policymakers do something to solve the problem. If we fail to avoid highly damaging climate change, the media will bear a lot of the blame.

5

The Reason behind the Strengthening Consensus: The Science Is Right

As concerns about man-made global warming and its potential consequences grew through the 1980s, the Intergovernmental Panel on Climate Change (IPCC) was formed. The IPCC was established by the United Nations Environment Programme and the World Meteorological Organization to provide the world with a clear scientific view on the current state of knowledge in climate change and its potential environmental and socioeconomic impacts.

The IPCC does not conduct any original scientific research or gather any data. Rather every five to seven years, the IPCC brings together the world's top climate scientists to review and assess the most up-to-date climate science research. The IPCC then produces a series of reports documenting the current state of our understanding of the climate, the risks it poses, and potential responses to those risks.

Thousands of climate science experts from all over the world contribute to the work of the IPCC on a voluntary basis. They are not paid for their contributions to the IPCC reports and put their own research on hold in order to contribute to the IPCC. Climate scientists with the greatest expertise in the subject of each chapter of the report are invited to be lead authors. Currently, 194 countries are members of the IPCC, and its reports are considered some of the best scientific documents in the world.

Each IPCC report has a Summary for Policymakers that represents a consensus of national representatives. In the early 1990s when the IPCC was producing its first reports, the consensus on human-caused global warming had already begun to form in the peer-reviewed climate science literature. It's also worth noting that contrary to the myth that the IPCC reports are "alarmist," because their summaries require a consensus among participants, they tend to err on the conservative side. It's difficult to achieve consensus among hundreds of participating countries, especially when some of those countries' economies rely heavily on fossil fuels.

1990

The IPCC First Assessment Report (FAR) was published in 1990. In its Working Group I report on the physical science basis, the FAR used climate models to project future global warming under various carbon dioxide emissions scenarios. Details about the climate models used by the IPCC are provided in Chapter 6.6 of the report.[1]

The modeled scenarios included business-as-usual (BAU) emissions and three other scenarios (labeled B, C, and D) in which global human greenhouse gas emissions began slowing in 2000. In 2010, the atmospheric carbon dioxide concentration in BAU projected by the FAR was approximately 400 ppm and in Scenarios B, C, and D was approximately 380 ppm. In reality, it was 390 ppm,[2] so we have thus far been right between the various scenarios considered in the IPCC FAR report.

Sulfate aerosols were still a major source of uncertainty in 1990. In an improvement over Kellogg's 1979 paper, the IPCC FAR correctly reported that an increase in atmospheric sulfate aerosols would cause an overall cooling effect. However, climate scientists at the time had difficulty quantifying this cooling effect (in fact, quantifying the aerosol cooling effect is still a major challenge for climate scientists today), and they also did not know how human aerosol emissions would change in the future.

The IPCC FAR ran climate simulations using models with equilibrium climate sensitivities of 1.5°C (2.7°F), 2.5°C (4.5°F), and 4.5°C (8.1°F) for a doubling of atmospheric carbon dioxide. However, in 1990, the IPCC overestimated the size of the energy imbalance created by a doubling of atmospheric carbon dioxide and thus overestimated the equilibrium climate sensitivity of each model. Using current estimates of the energy imbalance caused by a doubling of atmospheric carbon dioxide, the actual equilibrium climate sensitivity of the IPCC FAR "best estimate" model was actually 2.1°C (3.8°F) for doubled carbon dioxide, rather than the stated 2.5°C, for example. This is toward the lower end of today's best estimates of the climate sensitivity to the increased greenhouse effect (1.5 to 4.5°C, or 2.7 to 8.1°F, warming in response to doubled carbon dioxide levels in the atmosphere).

In order to evaluate how accurate the IPCC climate models were, we want to account for the difference between the greenhouse gas emissions scenarios modeled in the FAR report and the actual observed emissions. Figure 5.1 takes these changes into account and compares the IPCC best estimate climate model projections to the observed temperature change from 1990 to 2010. This is only a model of temperature changes in response to human greenhouse gas emissions and does not take natural climate effects into account.

As Figure 5.1 illustrates, the greenhouse gas–only model doesn't simulate the temperature change in the early 20th century well, but it does accurately project the global warming starting around 1950. Since the IPCC FAR only modeled temperature responses to greenhouse gas changes, this suggests that human effects began to drive global warming starting in the mid-20th century, while much of the early 20th-century warming was caused by natural effects like ocean cycles and an increase in solar activity.

The IPCC FAR best estimate climate model with just a 2.1°C (3.8°F) equilibrium climate sensitivity to doubled carbon dioxide matches the observed global warming since 1990 very accurately (predicting about 16 percent more warming than observed). However, while the model

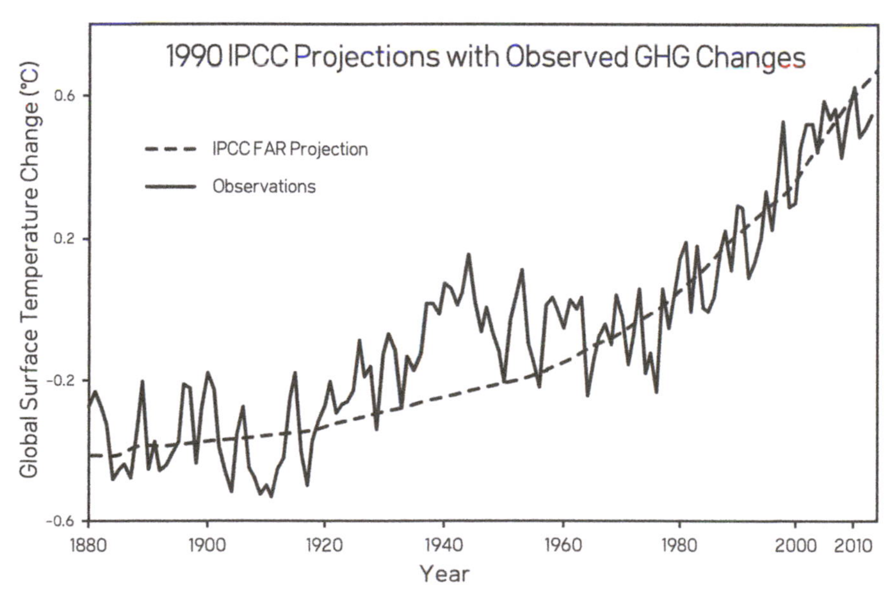

Figure 5.1 IPCC FAR "Best Estimate" Model Temperature Projections under Actual Greenhouse Gas Changes from 1880 to 2013 versus Observed Temperature

included the warming effects from not just carbon dioxide, but all human greenhouse gas emissions, it did not include the cooling effects from human sulfate aerosol emissions.

By including the main warming effects but excluding the main cooling effects, the IPCC FAR model overestimated the total energy imbalance caused by human emissions. Therefore, from this model we can expect the temperature change to a doubling of atmospheric carbon dioxide (the climate sensitivity) to most likely be somewhat higher than 2.1°C (3.8°F).

In any case, the accuracy of the 1990 IPCC global surface warming projections is impressive and provides evidence that carbon dioxide has played a dominant role in the global warming over the past two decades.

1995

The IPCC followed up the FAR with its Second Assessment Report (SAR) in 1995. In 1992, the IPCC published a supplementary report to the FAR,[3] which utilized updated greenhouse gas emissions scenarios called "IS92a" through "IS92f." The 1995 SAR continued the use of these IS92 greenhouse gas emissions scenarios and made comments regarding the advancement of our understanding and ability to model the global climate accurately.

> The increasing realism of simulations of current and past climate by coupled atmosphere-ocean climate models has increased our confidence in their use for projection of future climate change. Important uncertainties remain, but these have been taken into account in the full range of projections of global mean temperature and sea level change.

Perhaps one of the biggest improvements between the IPCC FAR and SAR was the increased understanding of and thus the ability to model the effects of sulfate aerosols. Section B.6 of the report discusses the subject.[4]

> Aerosols in the atmosphere influence the radiation balance of the Earth in two ways: (i) by scattering and absorbing radiation—the direct effect, and (ii) by modifying the optical properties, amount and lifetime of clouds—the indirect effect. Although some aerosols, such as soot, tend to warm the surface, the net climatic effect of anthropogenic aerosols is believed to be a negative radiative forcing, tending to cool the surface.

Of all the IS92 emissions scenarios considered by the IPCC, the IS92a and IS92b scenarios have been closest to actual human greenhouse gas emissions since 1995, with scenarios IS92e and IS92f running just a bit high. Scenarios IS92c and IS92d (which represent humans taking serious steps to reduce greenhouse gas emissions) increasingly diverge from reality, since we have not yet made serious efforts to reduce the human impact on the global climate. However, by 2015, the atmospheric carbon dioxide concentrations in each scenario are not very different.

One interesting aspect in the IS92 scenarios is that the IPCC projected the global energy imbalance caused by sulfate aerosols remaining strong throughout the 21st century. Given that aerosols have a short atmospheric lifetime of just one to two years (they wash out of the atmosphere relatively quickly), maintaining this strong influence would require maintaining human aerosol emissions (e.g., from burning coal and diesel) throughout the 21st century.

Because air quality and its impacts to human health are another concern related to sulfate emissions (e.g., China is currently working on reducing its sulfate pollution emissions to address its terrible air quality problems and the health impacts and social unrest that result), it's likely that human aerosol emissions will decline as the century progresses. This issue was one significant change made in the IPCC Third Assessment Report (TAR) projections, as we'll see later in this chapter.

The IPCC SAR also maintained the best estimate equilibrium climate sensitivity used in the FAR of 2.5°C (4.5°F) for a doubling of atmospheric carbon dioxide but, like the FAR, overestimated the global energy imbalance caused by this increase in carbon dioxide. Thus, as in the FAR, the actual equilibrium climate sensitivity of the FAR best estimate climate model was 2.1°C (3.8°F), which is again toward the lower end of today's best estimates.

Using the various IS92 emissions scenarios, the SAR projected the future average global surface temperature change to 2100. Each of the scenarios results in approximately the same amount of projected global warming between 1990 and 2013, so there's no need to adjust for projected versus observed greenhouse gas emissions in the comparison to the observed global warming (Figure 5.2).

The IPCC SAR projection is similar to the FAR projection shown in Figure 5.1. Both models have the same equilibrium climate sensitivity of 2.1°C (3.8°F) for a doubling of atmospheric carbon dioxide, but Figure 5.2 shows the difference when the cooling from human sulfate aerosol emissions is taken into account. Now the model has underestimated the global warming over the past two decades by about 20 percent.

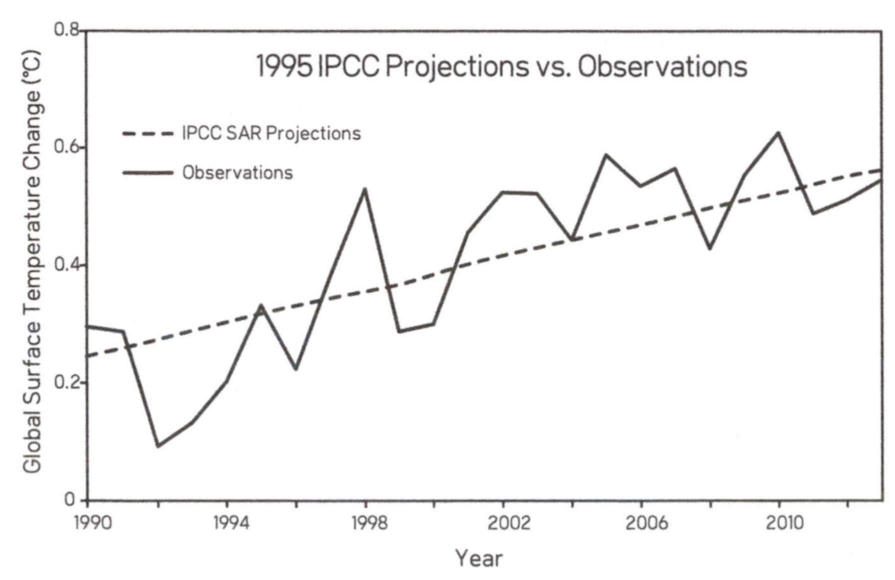

Figure 5.2 IPCC SAR "Best Estimate" Model Temperature Projections versus Observed Temperature

So much for the IPCC and mainstream climate scientists being alarmist.

1998

Geologist Don Easterbrook from Western Washington University was one of the first scientists "skeptical" about man-made global warming to make a concrete prediction about future global surface temperature changes. In 1998, Easterbrook predicted that the Earth would cool during the first 30 years of the 21st century, based on his interpretation of natural climate cycles. As Easterbrook put it,[5]

> The real question now is not trying to reduce atmospheric CO_2 as a means of stopping global warming, but rather (1) how can we best prepare to cope with the 30 years of global cooling that is coming, (2) how cold will it get, and (3) how can we cope with the cooling during a time of exponential population increase? In 1998 when I first predicted a 30-year cooling trend during the first part of this century, I used a very conservative estimate for the depth of cooling, i.e., the 30-years of global cooling that we experienced from ~1945 to 1977. However, also likely are several other possibilities (1) the much deeper cooling that occurred during the 1880 to ~1915 cool period, (2) the still deeper cooling that took place from about 1790 to 1820

during the Dalton sunspot minimum, and (3) the drastic cooling that occurred from 1650 to 1700 during the Maunder sunspot minimum.

You may recognize Easterbrook's references to the Dalton and Maunder grand solar minima discussed earlier in this book. Several recent papers have investigated the influence of these extended periods of low solar activity on global surface temperatures and found that another similar grand solar minimum would cause no more than 0.3°C (0.54°F) global surface cooling. This cooling would only be temporary until the solar minimum ended and in the meantime would only offset about a decade's worth of human-caused global warming. Despite being a geologist rather than a solar scientist, Easterbrook believes these solar minima had much larger impacts on global surface temperatures.

Easterbrook has continued to stand behind his global cooling predictions. These predictions are based on the assumption that the approximately 30-year transitions between warming and cooling periods in the recent past will continue in the future. The major problem is that this is not a physically based assumption. For example, Easterbrook doesn't investigate what caused the 1910–1940 warming (partly rising solar activity, and some human greenhouse gas emissions, ocean cycles, and a few other factors) or the slight 1940–1970 cooling (human sulfate aerosol emissions probably played a big role here). If we don't examine what caused these past climate changes, how do we know they will continue in the future?

This failure to understand the causes of past climate change is apparent in Easterbrook's global cooling predictions, which he bases on the strengths of various past periods of global cooling. He doesn't examine what caused these past cooling events, so he can't predict how strong the proposed future cooling will be. All he can do is assume that what happened during past climate changes will happen in the future. His predictions are just assumptions that past cooling events will be repeated in the future.

The problem is that what's happening now is different than what happened in the past. Humans weren't burning immense quantities of fossil fuels or releasing tens of billions of tons of carbon dioxide into the atmosphere every year a century ago, as we are now. Thus, we shouldn't expect future climate changes to look like past climate changes. Ignoring the human contribution to global warming, as Easterbrook does, will surely lead to an incorrect prediction. So how have his predictions stacked up so far?

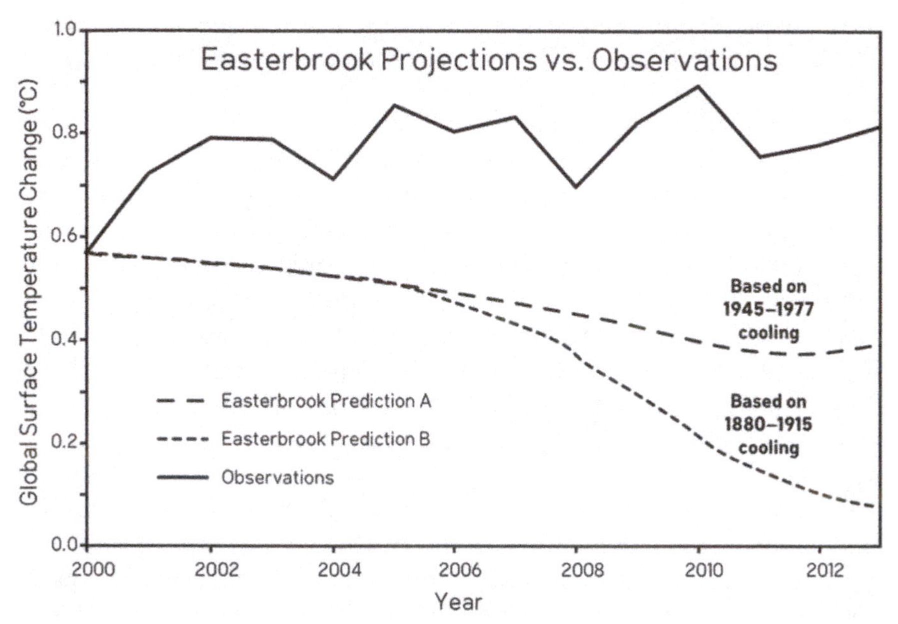

Figure 5.3 Easterbrook Global Cooling Prediction Based on 1945–1977 Cooling and 1880–1915 Cooling versus Observed Temperature

In short, over the first 13 years of his global cooling predictions, Easterbrook has already been wrong by between 0.34 and 0.68°C (0.61 to 1.2°F). As expected, this is a very poor result. For example, compare Easterbrook's projections to Wallace Broecker's 1975 prediction that we examined earlier, which was based mainly on the warming caused by human carbon dioxide emissions.

Broecker's 1975 prediction was within 0.3°C (0.54°F) of the observed global temperature in 2013, 38 years later, while Easterbrook's are off by 0.34 and 0.68°C (0.61 to 1.2°F) after just 13 years. This illustrates the importance of basing future predictions on solid physical footing (as opposed to ignoring human effects on the climate) and also shows that climate scientists understand the inner workings of the global climate much better than climate contrarians would have us believe. In fact, in the 1970s, climate scientists understood how the Earth's climate works better than many contrarian scientists do a full four decades later!

The inaccuracy of Easterbrook's predictions has not hurt his credibility in the conservative media, of course. In fact in January 2014, right-wing news outlet CNS News (essentially the Fox News of the Internet) ran a story about Easterbrook with the headline "Climate Scientist Who Got It Right Predicts 20 More Years of Global Cooling."[6]

The story was a remarkable example of outright denial of reality, claiming "Easterbrook's predictions were 'right on the money.'" The quotation marks in that sentence were particularly interesting, since the only person interviewed in the article was Don Easterbrook himself. Apparently Easterbrook told CNS News that his global cooling predictions were "right on the money," and rather than investigate those claims or even fact-check them with a real climate scientist, CNS News took his word for it. The story also quoted Easterbrook as saying,

> When we check [the IPCC] projections against what actually happened in that time interval [since 2000], they're not even close. They're off by a full degree in one decade, which is huge. That's more than the entire amount of warming we've had in the past century. So their models have failed just miserably, nowhere near close. And maybe it's luck, who knows, but mine have been right on the button.

None of the statements in that quote are even remotely close to being true. The IPCC global warming projections since 2000 have been quite accurate, to within about a tenth of a degree Celsius. Easterbrook is claiming they've somehow been off by a full degree; ten times larger than the actual discrepancy. Meanwhile Easterbrook's own predictions, which he claimed have been "right on the button" and "right on the money," have actually been 0.34 and 0.68°C (0.61 to 1.2°F) too cool. It's Easterbrook who's been off by nearly a full degree, not the IPCC.

In the same CNS News story, Easterbrook was also quoted attempting to dispute the expert consensus on human-caused global warming by citing a document called the Oregon Petition. This was an official-looking petition circulated by climate contrarians, claiming that there's no evidence that human-caused global warming will cause "catastrophic heating of the Earth's atmosphere and disruption of the Earth's climate" and that adding more carbon dioxide to the atmosphere would even be beneficial for plants and animals. The only requirement for signatories on the Oregon Petition is that they need to have some sort of college science degree. If you want to trust your climate science to petroleum geologists and lab technicians, the Oregon Petition is for you.

Those collecting the signatures also didn't check the claimed credentials of the signatories, so the first versions of the Oregon Petition listed many falsified names, for example, the names of members of the Spice Girls and several fictional characters from the television show M*A*S*H and the movie Star Wars.

Very few of the approximately 32,000 Oregon Petition signatories had any climate expertise, and many shouldn't even be considered scientists. For example, more than half of the signatories are engineers, medical professionals, computer scientists, and mathematicians. Removing those signatories from the list would reduce the number to closer to 13,000. The remaining signatories also represent only 0.1 percent of the scientists graduated in the United States over the past 40 years.[7]

In 2001, Scientific American attempted to contact a random sample of 30 of the signatories, finding,[8]

> Of the 26 we were able to identify in various databases, 11 said they still agreed with the petition—one was an active climate researcher, two others had relevant expertise, and eight signed based on an informal evaluation. Six said they would not sign the petition today, three did not remember any such petition, one had died, and five did not answer repeated messages. Crudely extrapolating, the petition supporters include a core of about 200 climate researchers—a respectable number, though rather a small fraction of the climatological community.

There are tens of thousands of climate researchers around the world, so 200 would represent somewhere in the ballpark of 1 percent of climate researchers. The Oregon Petition simply creates the false appearance that a large number of "scientists" dispute the threat posed by human-caused global warming by casting a wide net, allowing anybody with any college science degree to sign, no matter how irrelevant that degree or subsequent employment of the signatory is to climate science. In reality, the signatories on the Oregon Petition represent a tiny percentage of the general science and climate science communities. Most climate contrarians don't even bother to reference the Oregon Petition anymore, because it's now considered something of a joke, a desperate attempt to dispute the expert consensus on human-caused global warming.

Easterbrook made another similarly deceptive argument in the CNS News interview by claiming that carbon dioxide can't have a significant effect on the global climate because it comprises a small percentage of the gases in the Earth's atmosphere. This is a very similar ploy; the Oregon Petition is meant to cast doubt on the expert global warming consensus because the number of signatories *sounds* large. Talking about the percentage of the atmosphere comprised of carbon dioxide is meant to cast doubt on human-caused global warming because it *sounds* small. However, neither of these numbers means anything without the proper context. Thirty-two thousand "skeptic" signatories on a petition sounds large, until you realize that there are millions of Americans with science degrees who were eligible to sign the petition.

Likewise the 0.04 percent of the atmosphere comprised of carbon dioxide sounds small, until you realize that 99 percent of the atmosphere is comprised of non-greenhouse gases. The fact that there's lots of nitrogen and oxygen in the atmosphere doesn't tell us anything about the strength of the greenhouse effect from the carbon dioxide in the atmosphere. Moreover, there are chemicals that are toxic to humans when present in soil or groundwater in concentrations of just 0.000001 percent. Knowing only the concentration of a compound is not enough information to determine if it poses a threat to human or environmental health. Those who argue that carbon dioxide can't cause significant global warming because it comprises a small percentage of all gases in the atmosphere either are themselves not very smart or are banking on the ignorance of their audience.

In this case, Easterbrook's audience was CNS News writers and readers. He was probably right that the website's writers have a poor understanding of basic climate science. In particular, his claim that his predictions were on the money while those made by the IPCC were off by a degree were completely 100 percent backward. Yet when it comes to climate change, right-wing news outlets like CNS News don't care about the accuracy of the claims made by climate contrarians. As long as their comments support the news outlets' desired do-nothing message on climate change, they'll happily interview climate contrarians whose inaccurate predictions should have long ago discredited them.

Don Easterbrook is a perfect example of another climate contrarian with a history of being completely wrong on climate change, whose credibility nevertheless seems indestructible.

2001

The IPCC followed up the FAR and SAR with its TAR, published in 2001. Chapter 9 of the TAR discusses the report's projections of future climate change.[9]

> A few Atmosphere-Ocean General Circulation Model (AOGCM) simulations include the effects of ozone and/or indirect effects of aerosols (see Table 9.1 for details). Most integrations do not include the less dominant or less well understood forcings such as land-use

changes, mineral dust, black carbon, etc. (see Chapter 6). No AOGCM simulations include estimates of future changes in solar forcing or in volcanic aerosol concentrations.

There are many more AOGCM projections of future climate available than was the case for the IPCC Second Assessment Report.

In short, modeling of the effects of ozone and sulfate aerosols improved between the IPCC SAR and TAR, but some effects were still not well-simulated, like land-use changes (e.g., cutting down forests, which changes both the reflectivity of the Earth's surface and amount of carbon absorbed by the biosphere).

As noted earlier, the TAR reduced the projections of human sulfate aerosol emissions and their associated net cooling effect in the second half of the 21st century, which led to higher global warming projections by 2100. The IS92 emissions scenarios used in the SAR were replaced by the IPCC Special Report on Emission Scenarios (SRES), although the TAR also used the IS92 scenarios in order to allow comparisons with the projections in the SAR. The SRES cover a wide range of the main demographic, economic, and technological driving forces of future greenhouse gas and sulfur aerosol emissions. Each scenario represents a specific quantification of one of the four storylines described here.

- A1: a future world of very rapid economic growth, global population that peaks in mid-century and declines thereafter, and the rapid introduction of new and more efficient technologies. The three A1 groups are distinguished by their technological emphasis: fossil intensive (A1FI), nonfossil energy sources (A1T), or a balance across all sources (A1B).
- A2: a very heterogeneous world. The underlying theme is self-reliance and preservation of local identities. Fertility patterns across regions converge very slowly, which results in continuously increasing population. Economic development is primarily regionally oriented, and per capita economic growth and technological change are more fragmented and slower than in other storylines.
- B1: a convergent world with the same global population, that peaks in mid-century and declines thereafter, as in the A1 storyline, but with rapid change in economic structures toward a service and information economy, with reductions in material intensity and the introduction of clean and resource-efficient technologies. The emphasis is on global solutions to economic, social, and environmental sustainability, including improved equity, but without additional climate initiatives.

- B2: a world in which the emphasis is on local solutions to economic, social, and environmental sustainability. It is a world with continuously increasing global population, at a rate lower than A2, intermediate levels of economic development, and less rapid and more diverse technological change than in the B1 and A1 storylines. While the scenario is also oriented toward environmental protection and social equity, it focuses on local and regional levels.

The TAR ran these various emissions scenarios through global climate models to project possible future global temperature changes. As of 2015, we've so far been on track with the A2 emissions scenario.[10] Figure 5.4 compares the Scenario A2 global warming projections versus observed temperatures.

From this comparison, we arrive at a similar conclusion to that when we looked at the IPCC FAR projections. Up until the mid-20th century, most of the global surface temperature change was dictated by natural effects. Starting in the mid-20th century, human emissions became the dominant factor dictating the global temperature trend.

The IPCC TAR Scenario A2 projection is very close to the observed temperature change. Since 1990, the measured global surface temperature trend is 0.17°C per decade, while the IPCC TAR projected 0.16°C

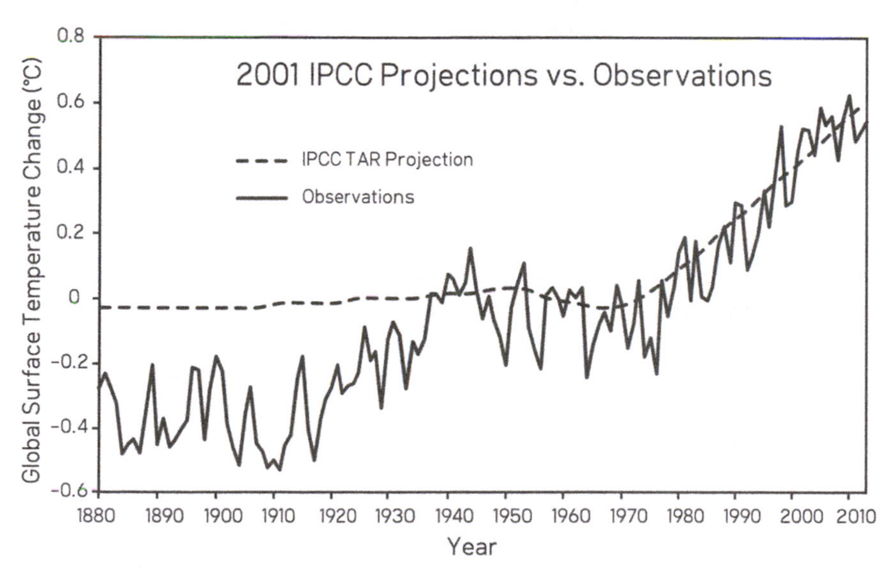

Figure 5.4 IPCC TAR "Best Estimate" Scenario A2 Model Temperature Projections versus Observed Temperature

per decade. The observations are also well within the envelope of all individual global climate model simulation runs.

It's important to note that if the observed temperatures don't precisely match the average of the model simulations, that doesn't make the models "wrong." The figures in this book depicting the IPCC projections only show the average of many individual model simulations, but each of those model runs represents a path that the climate could take. Hence, if the measured temperatures fall inside the "envelope" of individual model simulations, then the climate is behaving within the range of climate model expectations.

If we remain on track with the SRES A2 emissions scenario, the TAR projects that the average global surface temperature in 2100 will be approximately 4°C (7°F) above preindustrial levels, an exceptionally dangerous amount of global warming. The average effective equilibrium climate sensitivity of the climate models used in the IPCC TAR was 2.8°C (5°F) for doubled atmospheric carbon dioxide. Thus, the comparison between observations and the IPCC TAR-projected warming provides yet another piece to the long list of evidence that real-world equilibrium climate sensitivity is approximately 3°C (5.4°F) for doubled carbon dioxide.

2007

The IPCC followed up the FAR, SAR, and TAR with its Fourth Assessment Report (AR4), published in 2007. In its Working Group I (the physical basis) report,[11] Chapter 8 was devoted to climate models and their evaluation, and the report's Frequently Asked Questions discussed the reliability of models in projecting future climate changes. Among the reasons it cited that we can be confident in model projections is their ability to model past climate changes in a process known as "hindcasting." Hindcasting involves using models to simulate past climate changes and seeing how well their output matches the observational data.

> Models have been used to simulate ancient climates, such as the warm mid-Holocene of 6,000 years ago or the last glacial maximum of 21,000 years ago.

Global average surface temperatures during the last glacial maximum (ice age) were about 5°C (9°F) colder than today. It's worth noting that the average global temperature change from the peak of an ice age to the current warm period, which took many thousands of years, is similar to the amount of warming humans are on track to cause by 2100 if we continue on our current path. That's a bit of a scary thought.

The IPCC AR4 used the same SRES emissions scenarios as the TAR to project future temperature changes. As previously noted, we're currently on track with Scenario A2 emissions. Figure 5.5 compares the multi-model average for Scenario A2 to the observed average global surface temperature.

The global warming trend since 2000 is 0.18°C (0.32°F) per decade for the IPCC model average versus the observed 0.10°C (0.18°F) per decade during that time. The data falls within the model envelope and uncertainty range, but the observed trend over the past decade is lower than the average projection because we're considering such a short period of time. During that time, natural factors have acted to slow human-caused global warming, such as declining solar activity and an abundance of La Niña events in the Pacific Ocean.[12]

The IPCC AR4 was published only a few years ago, and thus, it's difficult to evaluate the accuracy of its projections at this point. We will have to wait another decade or so to determine whether the models in the AR4 projected the ensuing global warming as accurately as those in the FAR, SAR, and TAR.

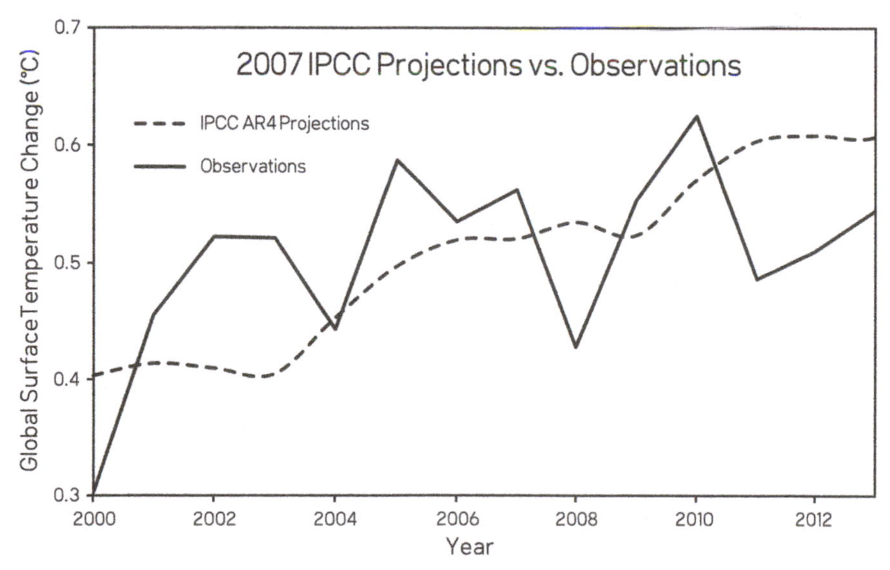

Figure 5.5 IPCC AR4 "Best Estimate" Scenario A2 Model Temperature Projections since 1990 versus Observed Temperature

Section 10.5.2 of the AR4 physical science report discusses the sensitivity of climate models to increasing atmospheric carbon dioxide and notes that the average climate sensitivity of the models used in the IPCC AR4 is 3.26°C (5.87°F) for a doubling of atmospheric carbon dioxide.

2009

In 2009, Syun-Ichi Akasofu, a geophysicist and director of the International Arctic Research Center at the University of Alaska-Fairbanks, released a paper arguing that the recent global warming is due to two factors: a "natural recovery" from the Little Ice Age (LIA) and "the multi-decadal oscillation" (which essentially refers to ocean cycles).[13] The LIA was a period of naturally declining global temperatures between the 16th and 19th centuries.

Akasofu argued that the current global warming trend, which he estimates at approximately 0.5°C (0.9°F) per century, actually began in the early 1800s and that

> this trend (0.5°C/100 years) should be subtracted from the temperature data during the last 100 years when estimating the manmade contribution to the present global warming trend. As a result, there is a possibility that only a small fraction of the present warming trend is attributable to the greenhouse effect resulting from human activities.

Akasofu also claimed that the multi-decadal oscillation can explain some of the warming over the past 35 years:

> This particular natural change had a positive rate of change of about 0.15°C/10 years from about 1975 (positive from 1910 to 1940, negative from 1940 to 1975), and is thought by the IPCC to be a sure sign of the greenhouse effect of CO_2. However, the positive trend from 1975 has stopped after 2000. One possibility of the halting is that after reaching a peak in 2000, the multi-decadal oscillation has begun to overwhelm the linear increase, causing the IPCC prediction to fail as early as the first decade of the 21st century.

Akasofu's hypothesis is essentially that there is a linear global warming trend caused by "recovery from the LIA," with natural climate oscillations superimposed upon it. In order for this to be a physically sound argument, Akasofu must explain the physical mechanism behind the LIA recovery and why this 0.5°C global warming trend continues to persist. What is the underlying cause?

One would expect a geophysicist like Akasofu to examine this question. After all, physics is all about figuring out what causes the physical world to behave the way it does. Unfortunately, nowhere in the 55 pages of his paper did Akasofu examine the physical cause of his purported 0.5°C per century warming trend since 1825. Most of the paper was spent looking at various regional temperature measurements, as well as data from ice cores, to show that the purported warming trend exists.

In a version of the paper published by an obscure journal in 2010,[14] Akasofu devoted a section to a discussion about galactic cosmic rays (high-energy particles from space which have a hypothesized, but unproven, and likely very small effect on the Earth's climate[15]), but did not attempt to quantify their effect. In fact, he began this section of his 2011 paper by stating:

> It is not the purpose of this section to discuss any major causes of climate change.

Instead, it appears that Akasofu assumed that the planet will naturally revert back to its previous state after a significant climate change as in the LIA. However, research by climate scientists has determined that the planet doesn't behave in the manner Akasofu suggests, simply "recovering" to some average natural state without some external force causing it to change. A paper published in the journal *Climatic Change* in 2011 concluded:[16]

> Temperature time series are not mean reverting. There is no evidence to support the idea that the observed rise in global temperatures are a natural fluctuation which will reverse in the near future.

Not only did Akasofu fail to examine the physical causes of the warming since the LIA, but he also failed to consider the possibility that a number of different factors are at play. For example, as discussed earlier in this book, increased solar activity, ocean cycles, and low volcanic activity contributed to the early 20th-century warming, but these natural factors have not contributed significantly to the warming since the mid-20th century. It is a logical failure to assume that a warming trend over nearly two centuries must have the same physical cause throughout the 200 years, and this argument is contradicted by the observational data (e.g., increasing solar activity in the early 1900s, but a slight decrease since mid-century).

A further failure of Akasofu's analysis is that while the linear warming trend over the past two centuries is approximately 0.5°C per century, nearly all of that warming has occurred over the past 100 years. In fact, most of the warming has happened over the past 40 years. Akasofu also failed to justify his assumption of a linear warming trend over the past two centuries. A slight warming in the 1800s, followed by faster warming in the early 1900s, followed by even faster warming over the past few decades—sounds rather like an accelerating trend, doesn't it? If you're going to fit a certain trend to the data, you first need a physical justification. What's the cause? Akasofu does not provide this justification, and without a physical reason, the choice of statistical trend fits is essentially arbitrary. Thus, Akasofu's entire premise is faulty on many different levels: physical, logical, and statistical.

Akasofu did discuss the cause of some of the variations in global temperature, with what he refers to as "the multi-decadal oscillation." The Pacific Decadal Oscillation, which consists in part of El Niño and La Niña cycles, is a primary component of these multi-decadal oscillations. These oscillations represent oceanic cycles that move heat from the oceans to air, and vice versa. However, these cycles just move heat around; they don't create it or store it. Thus, while they can cause significant short-term global surface temperature changes, they don't cause long-term global warming trends.[17] That's why they're called "oscillations" and represent the wiggles in Akasofu's model.

The only explanation for the long-term global warming provided by Akasofu is this unphysical concept of a recovery from the LIA. However, as discussed earlier, the Earth's climate doesn't just magically "recover" after a temperature change. Something has to force it to change, and Akasofu provided no physical explanation for his recovery.

Nevertheless, Akasofu did assume that the supposed 0.5°C (0.9°F) per century natural recovery will continue through at least 2100. So although he doesn't have a physical explanation behind it, Akasofu did predict continued global warming. Figure 5.6 compares Akasofu's global temperature model and predictions to the observed data since 2000.

As you can see, Akasofu predicted a very slight cooling (approximately 0.055°C, or 0.10°F) since 2000. Measurements show the Earth's surface warmed approximately 0.15°C (0.27°F) over that period,

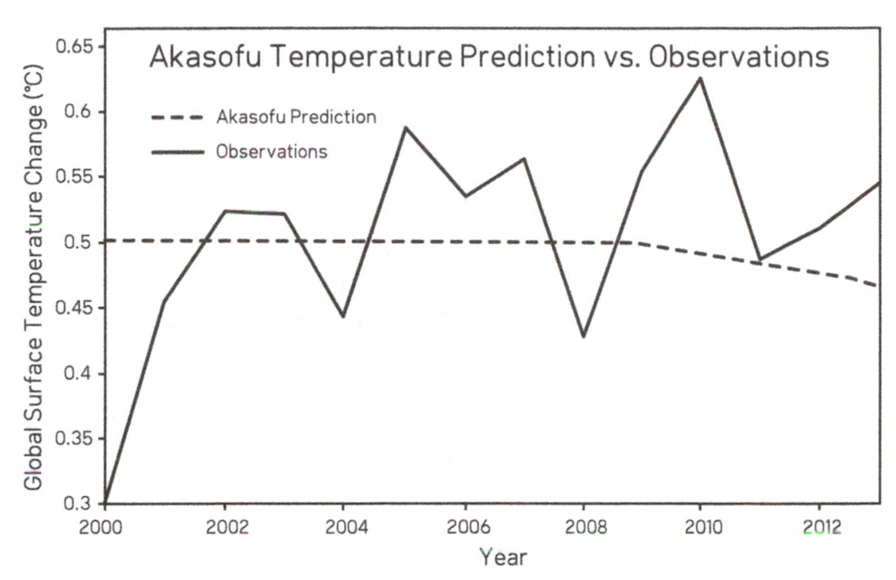

Figure 5.6 Akasofu Global Temperature Model and Prediction since 2000 versus Observed Temperature

despite natural influences acting to slow human-caused global warming over the past 15 years. Akasofu's prediction has not been terribly inaccurate yet, because it was made relatively recently.

However, given the expected atmospheric carbon dioxide increase over the 21st century, in order for Akasofu's predicted 0.5°C (0.9°F) per century global warming trend to hold true, the Earth's climate sensitivity would have to be in the range of 0.5 to 1.5°C (0.9 to 2.7°F) for doubled carbon dioxide, depending on how rapidly carbon dioxide continues to increase. This is below the range of likely climate sensitivity values, according to the IPCC and nearly all mainstream climate scientists. Thus, Akasofu's prediction is likely to underestimate future global warming.

It's also important to note that as in fellow climate contrarian Don Easterbrook's temperature predictions, Akasofu has completely ignored the warming effects of increasing atmospheric carbon dioxide in his predictions, assuming that whatever caused the preindustrial warming is also causing the current warming. Thus, Akasofu is really arguing that the climate sensitivity to carbon dioxide is effectively zero and that the observed and projected warming is due to some other "natural" effect that he has not identified. This is not a physical argument; Akasofu implies that extra heating from carbon dioxide (which we've directly measured with satellites) isn't causing any warming.

For Akasofu to be correct, this heat must somehow magically disappear. So even if his fellow climate change "skeptics" are somehow correct about low climate sensitivity and Akasofu's prediction turns out to be accurate, it will simply be due to sheer luck.

The argument made by Akasofu and other climate change contrarians that the Earth is warming only because it's "recovering from the LIA" is simply not a physical argument. Consider the analogy of climbing up and then falling down a hill. Akasofu's argument is akin to saying that you fall down the hill because you're just recovering from the increase in elevation from the climb; it makes no sense. You fall down the hill due to the force from the Earth's gravitational pull. The Earth's climate operates in the same way. It changes over the long term only when it's forced to change.

However, it is possible that the same external factors that caused the planet to cool during the LIA have subsequently caused global warming. Akasofu did not investigate this possibility, but we can. What caused the LIA?

Climate scientists believe a number of factors contributed to the LIA cooling. A decrease in solar radiation reaching Earth was certainly one contributor, as the LIA saw three distinct periods of low solar activity called the Spörer Minimum (1460–1550), Maunder Minimum (1645–1715), and Dalton Minimum (1790–1830). Solar activity has increased on average since the end of the LIA, but has remained flat over the past 50 years. Thus, while the sun was responsible for some of the early 20th-century warming, it cannot be responsible for the rapid global warming over the past half century.

The Earth also experienced heightened volcanic activity throughout the LIA.[18] Volcanic eruptions release sulfate aerosols into the atmosphere, which, as discussed in earlier chapters, block sunlight and cause the planet to cool. However, volcanic activity has also had a slight net cooling effect over the past century, particularly since 1950, and thus cannot explain the global warming over this period.

Another proposed contribution to the LIA cooling is a slowdown of the thermohaline circulation of the world's oceans through an introduction of a large amount of freshwater into the North Atlantic Ocean, potentially as a result of melting ice from Greenland. The Gulf Stream is part of the thermohaline circulation and transports warm water from the equator poleward toward Europe. If the North Atlantic Ocean becomes diluted with fresh water, this current could potentially become slowed or even shut down entirely. Wallace Broecker proposed this mechanism as a possible contributor to the LIA cooling.[19]

However, since the Greenland Ice Sheet has been shrinking rapidly due to the global warming over the past century, and the slowdown and potential shutdown of the thermohaline circulation has become a concern as a result, quite obviously the ocean conveyor has not had a warming effect over the past century.

Another interesting proposed cause of the LIA cooling involves humans. The Black Death caused a decrease in the human populations of Europe, East Asia, and the Middle East during the 14th century and a consequent decline in agricultural activity. A similar effect occurred in North America after European contact in the 16th century. Climate scientist William Ruddiman suggests that reforestation took place as a result of this reduced human population and agricultural activity, allowing more carbon dioxide uptake from the atmosphere to the biosphere, thus contributing to global cooling due to a decreased greenhouse effect during these periods.[20]

In this case, the exact opposite has indeed happened over the past century. The human global population has grown, as have deforestation and fossil fuel combustion, and thus the amount of carbon dioxide in the atmosphere. The resulting increased greenhouse effect has unquestionably contributed to global warming. However, I don't think human-caused global warming is what Akasofu had in mind when he described a recovery from the LIA. In short, there is simply no basis to the LIA recovery argument.[21]

In May 2013, a brand new peer-reviewed scientific journal called *Climate* published a paper by Akasofu that recycled his same old debunked LIA recovery argument in its very first edition.[22] It's difficult to know exactly how this happened, for example, whether one of the journal editors was friendly with Akasofu or whether the journal simply didn't find any qualified expert referees to review the paper. Whatever the reason, no peer-reviewed climate journal should have published Akasofu's unphysical arguments, especially given that the paper was nothing more than a recycling of his previously published papers from four years earlier. The publication of this poor paper caused a great deal of concern in the new journal's editorial staff, and one editor (Dr. Chris Brierley of the University College London) went as far as to immediately resign his editorial position. Brierley explained the reason behind his resignation:

> I do not believe that the paper is of sufficient quality for publication and have decided that I do not want to be associated with a journal with such lapses of judgment . . . the journal does not hold the

> standards that I feel should be strived for in science, leading to my resignation from the editorial board.[23]

Along with my colleagues John Abraham, Rasmus Benestad, and Scott Mandia, I put together a paper critiquing Akasofu's arguments and submitted it to *Climate*. In September 2013, the journal published our critique, which pointed out all the unphysical flaws in Akasofu's arguments that I have outlined in this chapter.[24] I applaud the efforts of the journal staff to correct the mistake they made in publishing Akasofu's flawed paper by quickly publishing our critique of it.

Any temperature prediction like Easterbrook's and Akasofu's that completely ignores the warming effects of carbon dioxide is fundamentally physically incorrect. Akasofu assumed a linear trend of unknown cause, an unknown periodic variability, and assumed that these two unknown phenomena will continue in the future, while disregarding what we know about the physics of the climate system.

Akasofu and Easterbrook both effectively threw out the physics established by Fourier, Tyndall, and Arrhenius over a century ago.

2011

A few more climate contrarians have finally begun to step up and make global temperature predictions of their own in recent years. In 2011, a rather obscure journal called the *Bentham Open Atmospheric Science Journal* published a paper by contrarians Craig Loehle and Nicola Scafetta.[25] Loehle is principal scientist at the National Council for Air and Stream Improvement, and his degrees are in forest science and range management. Scafetta is a research scientist in the Duke University physics department, where he mainly researches solar activity.

In their paper, Loehle and Scafetta attempted to model the global climate with a very simple formula consisting of three components: a 20-year natural cycle, a 60-year natural cycle, and a linear warming trend. In their paper, they tweaked the parameters in their simple model to match the observational global temperature data over the past 150 years and then used the best fit parameters to predict how global temperatures will change in the future.

Loehle and Scafetta were able to make their model fit the global temperature data fairly well from about 1850 to 1950. This isn't terribly surprising; as previously discussed, the human contribution to global warming didn't really start to kick in until the mid-20th century. After fitting their model to the 1850–1950 temperature data, Loehle and

Scafetta added in a second linear warming trend from 1950 to 2010, representing human influences on the climate.

There are a number of major flaws in the approach taken by Loehle and Scafetta in this paper. First, they did not place any physical constraints on the parameters in their model. For example, we know the energy imbalance caused by a doubling of atmospheric carbon dioxide is between 3.3 and 4.1 Watts per square meter, so a model should allow values for this parameter only within that range. A model that uses 10 Watts per square meter for this energy imbalance, for example, would not be physically realistic and thus would not accurately simulate the real world. It's critical to constrain physical parameters to a realistic range in order to accurately simulate the real world. Loehle and Scafetta did not do this.

Loehle and Scafetta also suggested that the 60-year cycle in their models was associated with astronomical cycles (they did not attempt to explain the source of their 20-year cycles):

> The solar system oscillates with a 60-year cycle due to the Jupiter/Saturn three-synodic cycle and to a Jupiter/Saturn beat tidal cycle.

The first obvious question to ask here is, why should cycles associated with Jupiter and Saturn impact the Earth's climate? There's no reason to believe they should. In fact, blaming Earth's climate changes on astronomical cycles treads closer to astrology than science, which is why some people have begun to describe this sort of argument as "climastrology." If there is no physical mechanism by which the proposed effects can influence the Earth's climate, then there is no way to come up with realistic physical constraints on the climastrology model parameters.

This explains why Loehle and Scafetta allowed their parameters to vary freely, but this sort of exercise (fitting a graph with a model with unconstrained parameters) is known as "curve fitting." Climate scientist Raymond Pierrehumbert has also referred to it as "cooking a graph,"[26] because with enough free parameters, any statistical model can be made to fit any data. As the famous mathematician John von Neumann said about this sort of curve fitting,

> With four parameters I can fit an elephant, and with five I can make him wiggle his trunk.[27]

Another problem with the Loehle and Scafetta approach is that they didn't explain why we should expect their model to accurately predict future changes. This will be the case only if, assuming their model is correct, human-caused global warming continues at the same rate in the future as it has for the past few decades. However, if we continue on our BAU path, human carbon dioxide emissions and the global warming they cause will accelerate.

There's also no reason to believe a model can predict what will happen in the future if it can't accurately simulate the past. Loehle and Scafetta did not test the accuracy of their model in matching global temperature changes prior to 1850, but this is not hard to do. As a matter of fact, in a previous paper, Loehle himself created a reconstruction of global temperatures going back 2,000 years.[28] Another group of scientists led by Anders Moberg created one of the most highly regarded temperature reconstruction of the past 2,000 years,[29] which we can also use for comparison (Figure 5.7).

The Loehle and Scafetta model matches the reconstructed temperature trends reasonably well back through the LIA, but fails miserably to match temperatures more than 500 years ago. Moreover, the 60-year cycle in their model matches up extremely poorly with the Moberg reconstruction, and even with Loehle's own reconstruction.

Several times between 1500 and 1900, the Loehle and Scafetta model is out of phase with both reconstructions, with the peak of the 60-year cycle coming at the same time as a trough in temperature. Thus, we see

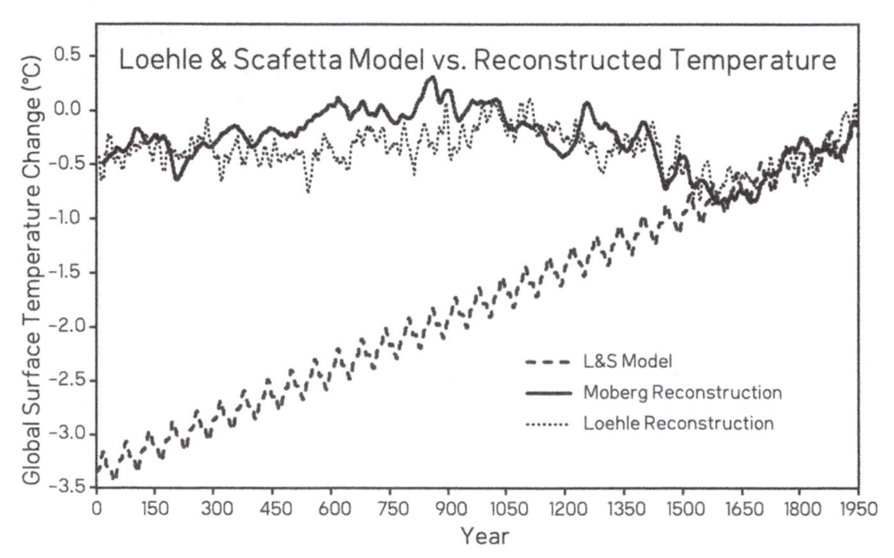

Figure 5.7 The Loehle and Scafetta Model Projected Back in Time Compared to the Loehle and Moberg Millennial Temperature Reconstructions

that although Loehle and Scafetta have gotten lucky and matched the temperature trend a few centuries into the past, the 60-year astronomical cycle that forms the basis of their paper is nowhere to be found in the temperature data.

There's yet another problem in this paper. Loehle and Scafetta describe the linear warming trend in their model from 1850 to 1950:

> The linear trend would approximately extrapolate a natural warming trend due to solar and volcano effects that is known to have occurred since the Little Ice Age.

The problem here is that in order to create a linear warming trend, the energy imbalance caused by these effects must be roughly constant during the whole period in question. However, Loehle and Scafetta's model applies this natural linear warming trend not only from 1850 to 1950 but also to 2010, and in their future global warming predictions. In short, for their model and future predictions to be correct, solar and volcanic effects must have an ever-increasing warming effect from 1850 to 2100 (when their global warming prediction ends).

However, as previously noted, solar activity hasn't increased over the past six decades, and volcanic eruptions have had a cooling effect over that period. Thus, by assuming that these two effects would cause a linear warming trend during a period when we know they actually had a slight cooling effect on global temperatures, Loehle and Scafetta have badly underestimated the human-caused global warming effect. Since there's no reason to assume solar and volcanic effects will cause constant warming over the next century, this undermines their future global temperature predictions even further.

Since Loehle and Scafetta just published their paper a few years ago, we can't yet evaluate the accuracy of their future predictions, which call for a global surface warming of 0.6°C (1.1°F) between 2000 and 2100. However, we know that because their model has no physical basis and does not accurately reproduce past temperature changes, there is no reason to put any faith in its future temperature predictions. We also know that they have underestimated the human-caused global surface warming over the past century and thus will underestimate the global warming over the next century as well.

The curve fitting mistake made by Loehle and Scafetta is a common one and can also be described as confusing correlation with causation. It occurs when an individual sees that two sets of data that appear to have similar changes (correlations) and stretches and manipulates one parameter until it matches the other. Then, once he has made them match, he declares that the first parameter is causing the changes in the second. For example, swimsuits have gotten smaller over the past century as the planet has warmed. I can make the swimsuit size data match the global temperature data, but that doesn't mean the Earth is warming because we're using less fabric to make swimsuits.

My colleagues and I came upon another example of this style of curve fitting in a paper by the University of Waterloo's Qing-Bin Lu, published in a physics journal in 2013.[30] The paper tried to argue that global warming was being caused not by human carbon dioxide emissions, but rather by our chlorofluorocarbon (CFC) emissions. CFCs are the chemicals that caused ozone depletion and the hole in the ozone layer, and whose use we began to phase out in the early 1990s after the international Montreal Protocol agreement was signed in order to address the environmental and health threats posed by ozone depletion.

Lu observed that recent average global surface temperature data seemed to correlate more closely with changes in human CFC emissions than with carbon dioxide emissions. That apparent correlation turns out to be due in large part to a cool bias in the global surface temperature record, which will be discussed in chapter 6 of this book. In any case, scientists have measured the global energy imbalance caused by carbon dioxide and CFCs (which are also greenhouse gases), and the imbalance caused by carbon dioxide is much larger because there's much more of it in the atmosphere. That's a difficult fact to get around when trying to blame global warming on CFCs instead of carbon dioxide.

Lu tried to bypass this roadblock to his hypothesis by going back to the same mistake made by Knut Ångström over a century ago, discussed earlier in this book, in which Ångström argued that the greenhouse effect from carbon dioxide has become saturated. Unfortunately, that hypothesis has been disproved for many decades and continues to be disproved by measurements of incoming and outgoing radiation on Earth. Nevertheless, Lu used this incorrect and long-disproved argument to disregard the warming effects of carbon dioxide. Lu then engaged in a curve fitting exercise, stretching the CFC emissions data curve until it matched the global surface temperature data curve, leading him to incorrectly conclude that changes in CFCs were the actual cause of global warming.

The paper contained yet another fundamental flaw. If the global energy imbalance were truly decreasing due to a decrease in human

CFC emissions, then the amount of heat trapped on Earth should be decreasing. The warming of global surface temperatures represents only a small percentage of the total heat energy building up on Earth, most of which accumulates in the oceans. When we account for all of the heat in the entire global climate system, global warming hasn't slowed at all. This fact by itself also disproved Lu's hypothesis.

The paper was published in a fairly obscure physics journal (the *International Journal of Modern Physics B*) rather than a climate journal, suggesting that it likely was not reviewed by any climate experts prior to its publication. When a climate-related paper is published in a non-climate journal, that's always a red flag that the authors couldn't get it past peer-review by climate experts. Nevertheless, once it was published, Lu's paper received a substantial amount of press coverage. The University of Waterloo issued a press release wrongly declaring that "Lu's theory has been confirmed,"[31] and some science journalism organizations essentially just copied that press release, compounding the error and spreading it to a wider audience.[32]

Several of my colleagues at Skeptical Science and I decided that it would be worth the effort to submit a response to the journal detailing the fundamental mistakes in the Lu paper, in order to correct the record. We discussed the fact that observational data disproves the "carbon dioxide is saturated" argument, that the amount of heat building up in the global climate has not slowed and so Lu's correlation was not even accurate, that Lu also used an outdated and inaccurate reconstruction of solar activity to explain the global warming in the 20th century, and several other mistakes. Like many journals, the *International Journal of Modern Physics B* asks the authors to suggest reviewers when a paper is submitted, and I made sure to only suggest climate scientists with relevant expertise.

When we received the reviews of our submission, the comments were extremely positive. One reviewer wondered why the journal had published Lu's paper to begin with, given its clear fundamental errors, and praised our team for taking the time to submit a paper explaining those mistakes. There's not much glamor in publishing papers that merely try to replicate previous studies and point out where they've gone wrong. Nevertheless, it can sometimes be a worthwhile exercise, particularly when a fundamentally flawed paper has received attention in the mainstream media, thus misinforming a large number of people.

The journal published our critique of Lu's paper in April 2014.[33] Not surprisingly, our critique received far less media attention than

Lu's original paper, despite the fact that Lu's paper was fundamentally flawed and our response was scientifically sound. In fact, my own article in *The Guardian* was the only media story about our paper. It just goes to show that journalists are interested in controversial stories that seem to contradict what we know, regardless of the factual accuracy of those stories. "Another study shows that carbon dioxide is causing global warming" just doesn't make for a catchy headline.

THE "WORST GLOBAL TEMPERATURE PREDICTION AWARD" GOES TO . . .

By far the worst global temperature prediction I've ever encountered was made by John McLean, a data analyst and member of the International Climate Science Coalition (ICSC). The ICSC is a group of climate change contrarians who attempt to cast doubt on the man-made global warming theory and are funded by political think tanks. On March 9, 2011, McLean made a rather extreme prediction about short-term global cooling:[34]

> It is likely that 2011 will be the coolest year since 1956 or even earlier.

The reasoning behind McLean's prediction was similar to Akasofu's multi-decadal oscillation. In 2009, McLean was the lead author on a scientific paper along with two other climate change contrarians that found that the El Niño Southern Oscillation (ENSO, which consists of El Niño and La Niña cycles and is one of the multi-decadal oscillations Akasofu referred to) accounts for a significant percentage of the short-term variation in the temperature of the Earth's lower atmosphere (troposphere) and that there is a roughly seven-month delay before changes in ENSO are seen in tropospheric temperatures.[35]

This was not a terribly Earth-shattering or new finding. Several other papers have arrived at very similar conclusions about the short-term effects of ENSO on surface and atmospheric temperatures. However, the McLean paper managed to sneak an incorrect and unsubstantiated statement into their conclusions:

> Overall the results suggest that the Southern Oscillation exercises a consistently dominant influence on mean global temperature.

This is an untrue statement that was not supported by McLean's research. What McLean's paper found was that the short-term wiggles in the temperature of the lower atmosphere could largely be explained by changes in ENSO. However, this is not at all the same thing as having a dominant influence on the Earth's temperature. Over periods of a few years or decades, changes in ENSO can either dampen or amplify the global warming trend, but ultimately it is greenhouse gases that have a dominant influence on the long-term global temperature trend, which is why the Earth's average surface temperature has risen 0.8°C (1.4°F) over the past century. ENSO causes only short-term wiggles around that long-term upward warming trend. El Niño events cause global surface temperatures to warm, but that effect is offset when a La Niña event happens and causes surface temperatures to cool.

But back to McLean's prediction about 2011 temperatures, why did I describe it as "extreme"? Well, according to our best estimates, the average global surface temperature has increased about 0.7°C (1.3°F) since 1956. In order to match the temperature in 1956 (which happened to be a relatively cool year), the Earth's average surface temperature needed to cool 0.84°C (1.5°F) from 2010 to 2011. The largest single year-to-year global temperature change on record (over the past 150 years) is 0.32°C (0.58°F). In order for McLean's prediction to be correct, the cooling from 2010 to 2011 would have to be nearly three times greater than any previous year-to-year temperature change on record.

As the saying goes, a picture is worth a thousand words. Figure 5.8 shows just how extreme John McLean's prediction was.

Halfway through 2011, average global surface temperatures were on track to be 0.66°C (1.2°F) hotter than McLean had predicted. I encountered McLean commenting on an article on the Australian website *The Conversation*, which publishes articles written by academics and researchers. I confronted McLean with his prediction and the fact that the planet was not on its way to freezing over (which was the only way his prediction could turn out to be remotely accurate), to see if he would admit his error. On the contrary, McLean stood behind his initial prediction.[36]

> Last time I looked 2011 wasn't over yet. It's a bit premature of you to be crowing about an annual average when the year isn't complete.

Seeing an opportunity to cash in on his false bravado, I offered him a wager. If McLean's prediction were wrong by less than half a degree Celsius at the end of 2011, he would win. If it wound up being incorrect by a least half a degree Celsius, I would win. Given that the average global surface temperature has not changed by more than 0.32°C

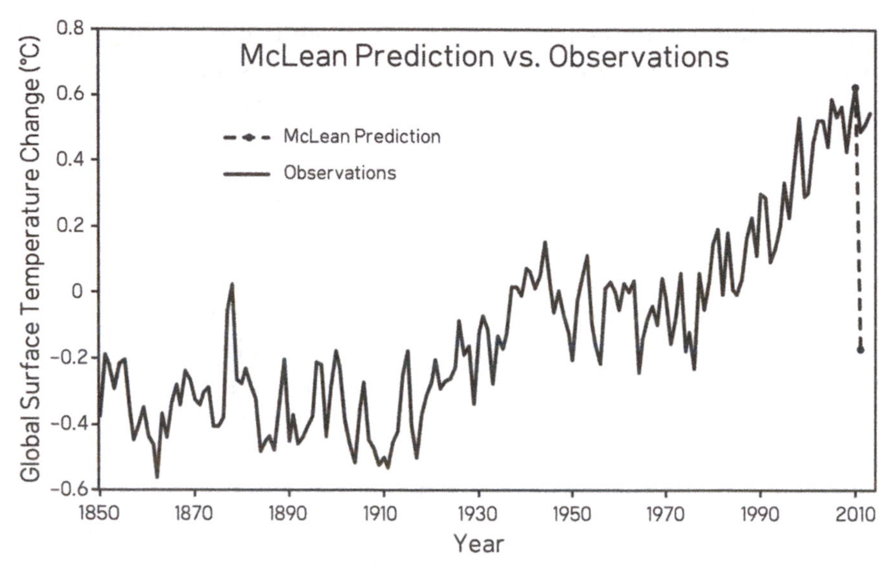

Figure 5.8 John McLean's 2011 Global Temperature Prediction versus Observed Temperature

from one year to the next, I thought that giving him a 0.5°C margin of error was a generous offer. Alas, McLean did not respond to my wager.

Of course, given that McLean's prediction was on track to be wrong by such a large margin, it was a wise decision not to take the bet. He turned out to be wrong by a whopping 0.66°C (1.2°F).

The failure of McLean's prediction proves that the unsubstantiated conclusion of his paper was wrong. The year 2011 was cooler than 2010 because the former was influenced by a La Niña cycle (which causes short-term cooling of surface temperatures), while the latter was influenced by an El Niño (which causes short-term warming on the Earth's surface). However, the difference between 2010 and 2011 was not gargantuan, as McLean predicted, because man-made global warming hasn't disappeared, and ENSO (and other ocean cycles) aren't really the dominant effect on global temperatures.

Given McLean's dramatic global cooling prediction failure, you might expect that his credibility on climate change would take a serious hit. If a mainstream climate scientist predicted a 0.8°C warming of average global surface temperatures over just a one-year time frame, you can bet that when that foolhardy and scientifically unsupportable prediction failed to come to fruition, climate contrarians would make sure that scientist's credibility was irreparably damaged.

On the contrary, in January 2014, *The Age* and *The Sydney Morning Herald* (both of which are normally respectable Australian periodicals) both published an opinion-editorial penned by John McLean.[37] The editorial was purely an attack on the IPCC with no supporting evidence. The credibility of the editorial was based entirely on the credibility of McLean, which *The Age* and *The Sydney Morning Herald* established by describing McLean as

> the author of three peer-reviewed papers on climate and an expert reviewer for the latest IPCC report. He is also a climate data analyst and a member of the International Climate Science Coalition.

We saw the quality of McLean's few peer-reviewed papers; they led him to make the worst-ever global temperature prediction. As for being an "expert reviewer" for the IPCC report, that requires nothing more than requesting to review the report drafts. There are thousands of expert reviewers who volunteer to review and make comments on the draft IPCC reports. Climate contrarians like McLean often sign up to be expert reviewers and make inaccurate comments that are rightfully disregarded by the IPCC authors. These contrarians thus have no meaningful impact on the final report. Nevertheless, they can proceed to claim to be IPCC expert reviewers. And of course the ICSC is just another climate contrarian group funded by political think tanks, so being an ICSC member is far from a sign of credibility.

The question is why *The Age* and *The Sydney Morning Herald* would publish an op-ed from John McLean to begin with, given the fact that his global cooling prediction was such a failure. Once again, it seems that climate contrarians can be spectacularly wrong about climate change, and somehow the media still treat them as credible experts on the subject.

6

The Continuation and Acceleration of Global Warming

Over the past few years, the most pervasive climate myth has been that global warming has "paused" or begun a "hiatus." In fact, this myth has appeared in virtually every climate contrarian mainstream media article, and many others seeking to present a "balanced" view of the subject (which have instead created false balance as a result).

The pause myth is based on the fact that the warming of global surface temperatures has slowed over the past 15 years or so. It's possible to carefully select starting and ending points to find a period over the past decade when the warming of surface temperatures has slowed down.

However, as I've shown in an animated graphic I created called "The Escalator,"[1] it's possible to find short periods of flat or cooling surface temperatures for 5- to 10-year periods during any decade since 1970. This is possible because short-term data is "noisy." There are many factors that influence short-term global surface temperature changes. For example, there's an 11-year solar cycle, volcanic eruptions, El Niño and La Niña events, other ocean cycles, and so forth.

Over the long-term, these influences all have close to zero effect on global surface temperature trends. Cycles, as the name suggests, are cyclical. They alternate between cool phases and warm phases, and over the long-term, these cancel each other out. Volcanic activity can have some longer-term temperature influences, if there's a period of relatively high or low volcanic activity, for example. However, particulates released from volcanic eruptions are washed out of the atmosphere fairly quickly, within a year or two. Thus over longer time frames, volcanic influences on global surface temperatures tend to be small, unless there's an extensive period of high or low volcanic activity.

However, it's possible for all of these influences to align in the same direction short term and temporarily cause a warm or cool period. For example, if the peak of the 11-year solar cycle, a few El Niño events, and a period of low volcanic activity all coincide, they would combine to cause a short-term warming period. If the opposite occurs with a period of low solar activity, more La Niña events, and higher volcanic activity, it would result in a short-term cooling period.

Over the past 10 to 15 years, we have indeed been in the midst of a period of relatively low solar activity and in an ocean cycle with a preponderance of La Niña events. With Chinese coal power consumption growing rapidly, human sulfate aerosol emissions and their associated cooling effect have probably grown over the past decade as well, and other influences like volcanic activity have also had a cooling effect on recent global surface temperatures.

The past decade has been something of a perfect storm, with all of these short-term influences on global surface temperatures acting in the cooling direction. Despite all these cooling influences, global surface temperatures have continued to warm slightly, or remain steady, depending on precisely what time frame we look at. This is a powerful illustration of the strength of human-caused global warming, that with all of these other influences acting in the cooling direction, temperatures have stubbornly continued to rise.

So how long can we expect the global surface warming slowdown to continue? That's a difficult question to answer. Solar activity is anticipated to remain relatively low in the coming decades. However, China is beginning to curb its coal consumption due to twin concerns about global warming and the health effects associated with its air pollution.

THE PAUSE THAT WASN'T

In 2013, Kevin Cowtan from the University of York and Robert Way of the University of Ottawa, both of whom are also my colleagues at Skeptical Science, published an important paper looking at another factor that has contributed to the apparent global surface warming slowdown.[2] As you might expect, in the Arctic region where there aren't many humans or much infrastructure in place, we don't have very many surface temperature monitoring stations. It also just so happens that the Arctic is the part of the planet that's warming fastest. This is due to feedbacks that amplify human-caused global warming in the Arctic region.

The main cause of this "Arctic amplification" is the loss of ice. There's a lot of ice floating on the Arctic Ocean, and ice is highly reflective, whereas the dark oceans absorb a lot more sunlight. As the Arctic sea ice has declined rapidly (losing about three-quarters of its total summer area and volume over the past three decades in what's been called the "Arctic sea ice death spiral"), revealing the dark oceans below, the Arctic has become much less reflective. Thus, as the region has warmed and the ice has melted, the Arctic has begun absorbing more heat from the sun, causing even further warming.

This presents a problem for scientists trying to measure the Earth's average surface temperature. Temperature measurements in the Arctic are particularly sparse, but that's the part of the planet that's warming fastest. For these reasons, climate scientists have long been aware that our estimates of the rate of warming at the Earth's surface are biased low. Some groups have tried to adjust the data to at least partially address this bias. For example, the scientists at NASA GISS use a weighted average between temperature stations, so that they can estimate surface temperatures where there are gaps in the measurements. This allows for a more complete global average temperature estimate than groups like the UK Met Office, who simply exclude the areas where measurements are lacking, thus creating a cool bias by missing the rapidly warming Arctic region.

In 2013, Cowtan and Way set out to find a better way to address this problem. A team of scientists with the Berkeley Earth Surface Temperature (BEST) project had recently used a statistical method known as "kriging" to interpolate between surface temperature measurements on land and shown that it was a very effective approach in accurately filling in the gaps between stations. Cowtan and Way applied this approach to both land and ocean surface temperature data, and they also tried a second approach, creating a hybrid of surface station and satellite measurements. Satellites measuring the temperature of the Earth's atmosphere have better coverage of the Arctic than surface temperature stations, so combining the two into a hybrid data set allowed Cowtan and Way to address the gaps in the data and create a second, independent global surface temperature data set for comparison.

They also used recently updated sea surface temperature (SST) data from the UK Met Office, which addressed some previous problems with SST data that have not yet been corrected in the data used by the NASA GISS team. So Cowtan and Way essentially put together all the best data and methodologies from several different scientific teams: the Met Office SST data, the kriging approach of the BEST team, and the satellite measurements put together by scientists at the University of Alabama at Huntsville (who happen to be among the few contrarian climate scientists who doubt humans are the main cause of global warming).

For the time period of 1997 through 2012, the UK Met Office data set had estimated the average global surface warming at 0.046°C (0.083°F) per decade, while the NASA GISS team estimated it at 0.080°C (0.14°F) per decade. That's a big difference, but as it turns out, both underestimated the actual warming. The new Cowtan and Way kriging and hybrid data sets estimated the global surface warming trend from 1997 to 2012 at 0.11 and 012°C (0.20 and 0.22°F) per decade, respectively. Overall, they found that the kriging method worked best to estimate temperatures over the oceans, while the hybrid method worked best over land and most importantly sea ice, which accounts for much of the unobserved region in the Arctic.

Nevertheless, both methods arrived at very similar results, estimating the actual global surface warming trend during that 16-year period at more than double the estimate of the widely used UK Met Office data set. The clever and well-executed approach of Cowtan and Way immediately received praise from many scientists with expertise in global temperature methods and relevant statistics, including scientists at the Met Office and NASA GISS.

It's worth noting that over the long term, the Cowtan and Way data sets are very similar to those of the Met Office and NASA GISS. It's only over the past 15 to 20 years that the Arctic amplification has really kicked into gear due to the rapid decline in Arctic sea ice, causing the significant short-term underestimate of surface temperatures among data sets omitting the Arctic region. Nevertheless, Cowtan and Way showed that the so-called global warming pause wasn't real and that the warming of global surface temperatures had slowed only by half as much as previously thought, though there was still some detectable slowing. Several other recent studies have suggested that while decreased solar activity and an increase in aerosol pollution in the atmosphere contributed to that slowed surface warming, the oceans have also played a big role by absorbing more heat, leaving less to warm the atmosphere.

GLOBAL WARMING IN THE OCEANS

A number of recent research papers have suggested that ocean cycles have played a major role in the recent slowing of global surface warming. In particular, what's known as the Pacific Decadal Oscillation (El Niño and La Niña events are part of this cycle) has been fingered as the prime suspect behind the slowed surface warming.

The challenge for climate models lies in the fact that at the moment, we can't predict ahead of time how these ocean cycles will behave. Climate modelers do try to include their influences on the global climate by including random ocean cycle variations in their models, but being unable to predict how these cycles will change in the future makes it difficult to accurately predict short-term global surface temperature changes.

A 2013 paper published in the journal *Nature* by Yu Kosaka and Shang-Ping Xie from the Scripps Institution of Oceanography[3] tried to get around this problem by using a global climate model, but instead of letting all the variables roam free, they constrained SST in the El Niño region in the tropical east Pacific Ocean to match observed historical data. Thus, rather than simply using random ocean cycle fluctuations, they trained their model to accurately simulate past changes in the Pacific El Niño cycle.

The resulting model run achieved a remarkably accurate simulation of past observed surface temperature changes. Despite the fact that the scientists constrained SSTs over an area comprising only about 8.2 percent of the whole globe, the simulations were able to accurately reproduce global temperature changes. And not only did they accurately reproduce the average global surface temperatures, but they also did well in simulating regional and even seasonal changes.

As it turns out, the global surface warming pause has happened only during the winter season in the Northern Hemisphere. This pattern is consistent with the simulations in the Kosaka and Xie model. Ocean current circulations that are responsible for heat transport are stronger in the winter than in the summer, and thus La Niña events in the Pacific have more influence on global surface temperatures in the Northern Hemisphere winter.

Perhaps most importantly, the Kosaka and Xie model also accurately simulated the slowed global surface warming over the past 15 years. As the authors wrote,

> Our results show that the current hiatus is part of natural climate variability, tied specifically to La-Niña-like decadal cooling. . . . For the recent decade, the decrease in tropical Pacific sea surface temperature has lowered the global temperature by about 0.15 degrees Celsius compared to the 1990s.

In July 2014, a similar paper was published in *Nature Climate Change* by Risbey, Lewandowsky, Langlais, Monselesan, O'Kane, and Oreskes.[4] Instead of forcing climate models to match Pacific SST observations, these scientists made use of a large set of existing simulations from 18 climate models. They looked at each 15-year period since the 1950s and compared how accurately each model simulation had represented El Niño and La Niña conditions during those 15 years, using the trends in what's known as the Niño3.4 index. Each individual climate model run has a random representation of these natural ocean cycles, so for every 15-year period, some of those simulations will have accurately represented the actual El Niño conditions just by chance. The authors concluded:

> When the phase of natural variability is taken into account, the model 15-year warming trends in CMIP5 projections well estimate the observed trends for all 15-year periods over the past half-century.

Thus, according to these studies, the majority of the slowed warming of global surface temperatures is due to ocean cycles, especially in the Pacific. Similarly, a paper published by Swiss climate scientists in August 2014 found that climate models could accurately reproduce the slowed surface warming when accounting for the surface temperature bias (quantified by Cowtan and Way), reduced solar activity, increased volcanic activity, and recent cooling from ocean cycles.[5] They estimated that from 1998 to 2012, ocean cycles caused about 0.06°C (0.11°F) global surface cooling, the sun caused 0.04°C (0.07°F), and volcanoes caused 0.035°C (0.063°F) cooling.

These results are also generally consistent with research led by Masahiro Watanabe of the Japanese Atmosphere and Ocean Research Institute. Watanabe's team published a paper in 2013[6] that ran simulations with various global climate models and found that they did not often accurately reproduce the recent slowed global surface warming.

The global energy imbalance from 2001 to 2010 in the climate models used by the Watanabe team was somewhat smaller than that in the observational data. Yet the climate models also simulated more surface warming than has been observed. This suggests that the slowed surface warming isn't primarily a result of a smaller global energy imbalance due to factors like increased cooling from human aerosol emissions and lowered solar activity. Instead, it suggests that internal variability in the Earth's climate system from factors like ocean cycles is more likely to be the main cause of the slowed global surface warming.

Similar to the Kosaka and Xie paper, the Watanabe team found a strong correlation between global surface temperatures and SSTs over the Pacific Ocean. They then applied a simple statistical correction using this relationship with SSTs to determine whether internal variability could explain the slowed global surface warming.

The Watanabe team found that indeed it could. The model simulated an enhanced, more efficient overall heat uptake by the oceans, which suggests that the slowed surface warming can be explained by internal variability transferring more heat to the deep oceans. This is likely a temporary effect, because climate models predict that the oceans will actually become less efficient at absorbing heat over time in a warming world. The Watanabe team concluded:

> Therefore, unless models miss effects of other forcing agents, it is likely that this [less efficient ocean heat uptake] process will occur and act to accelerate surface warming in coming decades.

ACCELERATED DEEP OCEAN WARMING

In fact, this expectation of more ocean warming is consistent both with the findings of previous climate modeling research and with observational data. Gerald Meehl of the National Oceanic and Atmospheric Administration led modeling research that resulted in papers published in 2011 and 2013 on this subject.[7,8] In these studies, Meehl and his colleagues found that in their model simulations, decade-long periods of little or no global surface warming were relatively common.

Meehl and his colleagues examined some of these hiatus periods and found that in decades where surface temperatures were stagnant, more heat than usual was being transferred to the deeper layers of the ocean, below 300 meters and particularly below 750 meters deep. The study also found that the general pattern of warming and ocean circulation in the model during these hiatus periods is very similar to that which occurs over shorter time frames during La Niña events.

In short, these studies also suggest that periods during which there is a preponderance of La Niña events in the Pacific (as has been the case over the past decade), more heat will be transferred to the deep ocean layers, which will act to slow the warming at the surface. However, when the Pacific cycle switches, the authors anticipate that surface temperatures will rise rapidly.

A paper published in 2013 led by Virginie Guemas of the Catalán Institute of Climate Science arrived at essentially the same conclusion.[9]

Similar to the approach taken by Kosaka and Xie, the Guemas team compared the short-term predictive ability of climate model simulations with and without information about the previous history of the observed ocean cycle changes. The question was whether accounting for past natural variability would allow the model to more accurately predict short-term future changes.

Like Kosaka and Xie, the Guemas study found that the model simulations that used past observed ocean cycle information were able to much more accurately forecast the recent slowed SST changes than the control run that didn't include this past ocean cycle data. Similar to the conclusions in the Meehl research, Guemas noted:

> If it is only related to natural variability then the rate of [surface] warming will increase soon.

The accelerated warming of the oceans isn't just a result in models and theories though; it's an observational reality. In the early 2000s, scientists began deploying a large collection of small, drifting oceanic robotic probes called the Argo network. By the end of 2007, the array included 3,000 floats around the world. Prior to the Argo network, ocean temperatures were measured by expendable and mechanical bathythermographs. A bathythermograph is an instrument that has a temperature sensor and is thrown overboard from ships to record pressure and temperature changes as it drops through the water.

Until recently, we didn't have very good estimates of the amount of heat absorbed by the deeper oceans, below about 700 meters. In 2012, a group of scientists from the National Oceanographic Data Center led by Sydney Levitus published a paper estimating the amount of heat absorbed by the oceans all the way down to 2,000 meters, and as far back as the year 1955.[10] Their study found that the oceans have been accumulating a whole lot of heat.

Specifically, over the past five decades, the oceans have absorbed over 100 trillion Joules of energy per second. That's the equivalent energy of more than two Hiroshima atomic bomb detonations per second since the 1950s, on average. The rate of ocean heat accumulation also appears to have accelerated. Since the early 1980s, the oceans have averaged the equivalent of three Hiroshima atomic bomb detonations per second. Since the late 1990s, we're up to four detonations per second.

In early 2012, the scientific journal *Physics Letters A* published a paper by Douglass and Knox, two climate contrarian professors from the University of Rochester, New York Department of Physics and Astronomy.[11] In that paper, they argued that the Earth goes through "climate shifts," meaning that it alternates between periods where it absorbs more heat and periods where it absorbs less.

There were several problems with the analysis in the paper. What caught my attention in particular was that Douglass and Knox claimed that since 2002, the climate has gone through one of its "cool shifts" and accumulated little heat. They also used this recent data to argue that the climate sensitivity to the increased greenhouse effect is low. Something was clearly amiss with their argument. For one thing, the amount of heat accumulated by the Earth's climate should be rising fairly steadily along with the increasing greenhouse effect. For another, the Earth's climate sensitivity should essentially be a constant number; it shouldn't rise and fall every few years when a new climate shift happens.

When I looked into their analysis more carefully, I saw that Douglass and Knox had used only ocean heat content data for the uppermost 700 meters of ocean. This wasn't surprising, because their paper was published just before the Levitus study that made ocean heat data to 2,000 meters in depth available to the public. However, Douglass and Knox were making claims about the overall global energy imbalance without considering the heat accumulating in the deeper ocean.

I decided to submit a comment to *Physics Letters A* along with some of my colleagues from Skeptical Science. We performed the same analysis as Douglass and Knox, but we used the Levitus ocean heat content data to a depth of 2,000 meters. One of the world's foremost oceanographers, John Church also joined us and provided land, atmosphere, and ice warming data he had used in a recent publication that also examined the accumulation of heat in the global climate.

When we put all the data together, we found that as expected, the overall heat building up in the Earth's climate had not slowed at all. In fact, more heat has accumulated over the past 15 years than the previous 15 years. Our paper was published later in 2012,[12] and its main result is shown in Figure 6.1.

Basically, what has happened is that while the warming of the surface and shallower oceans (to a depth of 700 meters) has slowed since 1998 (when we experienced one of the strongest El Niño events in the past century), the warming of the deep oceans has accelerated.

This was investigated further in a 2013 paper published in *Geophysical Research Letters* by Balmaseda, Trenberth, and Källén.[13] In their study, these scientists used ocean heat content estimates from

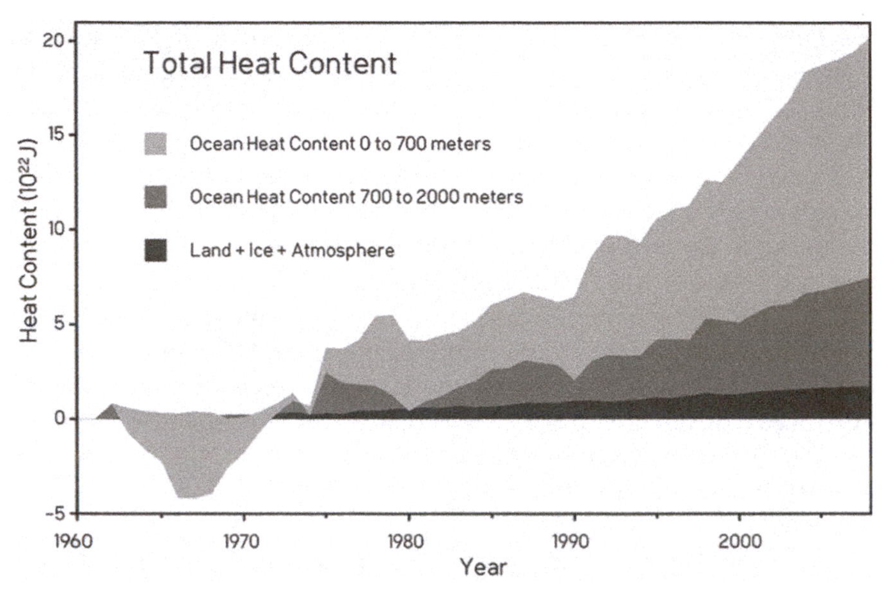

Figure 6.1 Land, Atmosphere, and Ice Heating, 0 to 700 Meter Ocean Heat Content Increase, 700 to 2,000 Meter Ocean Heat Content Increase

the European Centre for Medium-Range Weather Forecasts' Ocean Reanalysis System 4. A "reanalysis" is a climate or weather model simulation of the past that incorporates data from historical observations. In this case, the data incorporated into the model included ocean temperature measurements from bathythermographs and newer instruments on Argo buoys, and other types of data like sea level and surface temperatures.

Once they created their new estimate of ocean temperatures, the authors described the ocean warming since 1999 as

> the most sustained warming trend in this record of [ocean heat content]. Indeed, recent warming rates of the waters below 700m appear to be unprecedented. . . . In the last decade, about 30% of the warming has occurred below 700 m, contributing significantly to an acceleration of the warming trend.

This result is similar to the findings of our 2012 paper, in which we also found that about 30 percent of the warming over the past decade has occurred in the deeper ocean layers. Thus, the observational data is consistent with the model-based conclusions from Gerald Meehl's team, suggesting that during hiatus decades, while the warming of surface temperature slows, it's due to more heat being transferred to the deeper oceans. It's also consistent with the conclusions of the Watanabe team, suggesting that the oceans have recently become more efficient at transferring heat to the deep layers. The puzzle pieces were coming together to paint a consistent picture—global warming hadn't paused; instead more heat was temporarily going into the deeper oceans.

The study published by the team led by Balmaseda also found something curious in the ocean temperature data—the amount of heat being transferred to the deeper layers of the ocean since 2000 appears unprecedented, with deep ocean warming happening significantly faster than at any time in the past six decades. Subsequent research by Kevin Trenberth suggested that stronger ocean winds could account for this accelerated transfer of heat to the deep oceans. A paper published in the journal *Nature Climate Change* in February 2014, led by Matthew England from the University of New South Wales, added evidence to support Trenberth's theory.[14] England's paper found that trade winds in the Pacific Ocean have strengthened over the past two decades, to an unprecedented degree since our records begin (starting around the year 1900).

This strengthening of trade winds, caused in part by a phase of what's known as the Interdecadal Pacific Oscillation (IPO), has caused more heat from the surface to be mixed down into deeper ocean layers, while bringing cooler waters to the surface. The combination of these two processes also acts to cool global surface temperatures. Unprecedented trade wind acceleration has caused unprecedented warming of the deep oceans since 2000 and also acted to slow the warming at the Earth's surface.

However, all signs indicate that this is a temporary change. The thing to remember about cycles is that they're cyclical, and ocean oscillations like the IPO inevitably change phases. In the 1990s, the ocean cycles caused less heat to be transferred to the deep oceans and more to the shallow oceans and surface. Hence, global surface warming actually happened faster than the average global climate model expectation in the 1990s. Since 2000, the ocean cycles have switched, and surface warming has happened more slowly than the average model expectation.

Over the long term, these cycles average out. Hence, climate models do a good job of predicting long-term global surface temperature changes, but they struggle with short-term changes because ocean cycles are unpredictable. It's only a matter of time until the ocean cycles switch again and we enter another period of accelerated global surface warming.

This is all based on fairly cutting-edge recent scientific research. While the oceans appear to be the main culprit behind the slowed surface warming, other factors have likely played a role as well. For example, volcanic activity appears to have been above average over the past decade, solar activity is relatively low, and sulfate aerosols from Chinese coal power plants have risen.

The planet will warm in response to a global energy imbalance, so the question is how much of the slowed surface warming is in response to these factors decreasing the global energy imbalance, and how much is due to more of the warming being shifted to the deep oceans. This is still an open question being investigated by climate scientists, though it looks like about half of the surface warming slowdown was due to more heat being transferred to the oceans and half due to external factors like the sun and volcanoes.

However, humans have continued to increase the greenhouse effect by burning more and more fossil fuels. Basic physics tells us that the planet must warm in response to the energy imbalance caused by this increased greenhouse effect. Global warming simply can't magically stop; we have to reduce our greenhouse gas emissions to stop it. The warming at the surface can be temporarily slowed by more heat being transferred to the oceans, and by other factors offsetting some of the increased greenhouse effect, but these are only temporary changes. The only way to permanently slow or stop global warming is to slow or stop our greenhouse gas emissions.

There is a possibility that the unprecedented strengthening of the ocean trade winds and deep ocean heat storage may buy us a little time before the worst consequences of climate change strike. According to the study led by Matthew England, the changing phase of the IPO can account for only about half of the strengthening of the trade winds. Why the trade winds are now so much stronger than they were in previous similar IPO phases remains an open question that climate scientists continue to investigate. Climate scientist Michael Mann suggested in his book *The Hockey Stick and the Climate Wars* that changes in ocean cycles like this one may actually act as a negative feedback, dampening global warming. If some as of yet undetermined factor is causing ocean winds to become consistently stronger in a warmer world, resulting in a transfer of more global warming into the deep oceans, this could act to slightly slow the rate at which temperatures at the Earth's surface continue to warm. However, it remains to be seen whether this is an accurate description of what's happening, or whether the strengthening of the trade winds is a purely temporary change. In any case, there's a limit to how strong the trade winds can become, and to how much heat they can churn into the deep oceans, so it's unlikely that this effect could continue to significantly slow global warming at the Earth's surface.

THE MYTH OF THE PAUSE

Unfortunately, the media have struggled to keep up with this recent ocean research. Early reports did a good job documenting the warming of the deep oceans, but climate contrarians have managed to shift the focus onto the slowed global surface warming in a flood of opinion articles published primarily in politically conservative media outlets. Other media outlets began to follow suit, for worry of being perceived as biased for ignoring the opinions of "the other side." As a result, media coverage has suffered from false balance, where the minority view of climate contrarians is given a disproportionate amount of coverage.

Climate contrarians have insisted that the global surface warming pause is a problem for climate scientists and their models. They have declared that climate models didn't predict the slowdown ahead of time, and therefore climate scientists must not understand how the climate works. It's a classic logical fallacy, essentially arguing "if we don't know everything, we know nothing."

In reality, the long-term surface warming trend is consistent with global climate model projections, as illustrated throughout this book. In fact it's the climate contrarians whose predictions have failed. Nevertheless, by cherry-picking the data since the abnormally strong El Niño in 1998, the contrarians have successfully shifted much of the media climate focus onto the so-called pause.

It's true that climate models don't do well with projecting short-term changes, in large part because we can't yet predict ocean cycle changes ahead of time. But it's long-term climate changes that are important in terms of future climate and future generations. Global climate models do a good job in projecting those, because over longer periods of time the short-term cycles and noise cancel out, and the immense global energy imbalance caused by the increased greenhouse effect is the dominant cause of long-term global temperature changes. Climate predictions are the opposite of weather predictions; weather changes are harder to predict further in the future, but climate changes become easier to predict.

It's actually quite remarkable that global surface temperatures have continued to warm in recent years despite all the factors acting to cool them. Over the past two decades, solar activity has been low; volcanic activity and human coal burning have added aerosol particulates into the atmosphere that cause cooling by blocking sunlight; and the unprecedented trade wind strengthening has churned more heat into the oceans, leaving less to warm the atmosphere. Despite all these cooling effects, although the rate at which surface temperatures have risen has slowed, they have continued to rise. This illustrates the strong influence of human-caused global warming on surface temperatures.

THE 2014 IPCC REPORT

In 2014, the IPCC released its Fifth Assessment Report (AR5). Unfortunately, a draft version of the report was leaked nearly a year prior to the completion of the final report. This leak allowed those with an antiscience agenda and a lack of journalistic ethics to put their desired spin on the contents of the report ahead of its release. It was a situation ripe for climate contrarians to misinform the public, and misinform us they did.

As a prime example, the leaked draft of the report included a graph that compared the projected global surface warming from the climate models used in previous IPCC reports to the measured data, much like the figures presented in this book. However, the draft IPCC AR5 figure had a serious flaw. The models and observational data were aligned incorrectly, resulting in the false visual impression that surface temperatures had warmed more slowly than all the IPCC climate models had projected.

This was due to a mistake in a procedure called "baselining." Global average surface temperature data is not graphed in terms of absolute temperatures, but rather in temperature "anomalies," just like the graphs in this book. A temperature anomaly is the difference between the temperature at a given point and the average temperature during a particular reference period. For example, if the average temperature for the period 1951–1980 is 14°C (57.2°F) and the temperature in 2012 is 14.5°C (58.1°F), the temperature anomaly in 2012 is 0.5°C (0.9°F). Climate scientists plot temperatures in terms of anomalies because what we're really interested is the rate of change of global warming. It's easier for people to interpret the rate of change from 0 to 0.5°C (0 to 0.9°F) than from 14 to 14.5°C (57.2 to 58.1°F), for example.

The precise choice of baseline period isn't important, because it doesn't affect the rate of change. However, if baselining is done incorrectly in a graph, it can create a false visual illusion. That's what happened in the draft IPCC figure.

Normally, climate scientists use 30-year periods to baseline global surface temperature data, because over that amount of time, short-term noise averages out. Thus, using a 30-year baseline gives climate scientists a good reference for comparison of long-term trends. Instead, the draft IPCC report figure used only a single year (1990) as the baseline period.

The problem with that decision is that 1990 was an abnormally hot year—the hottest ever recorded up to that date in the global surface temperature measurements. But in the average of all model simulations, 1990 was just another normal year. Thus, using 1990 as the single baseline year for comparison resulted in aligning the temperature data low as compared to the model simulations. In effect, the temperature measurement data was shifted down compared to the average model simulations, and thus, it appeared as though the IPCC climate models had overestimated global warming.

The climate contrarians immediately jumped on this draft figure and announced that the IPCC had effectively admitted that its model simulations were failures. The story began in climate contrarian blogs but eventually seeped into more mainstream politically conservative media sources like Fox News.[15] However, climate realists like a blogger and statistician who goes by the pseudonym "Tamino" immediately recognized and documented the baselining error in the draft IPCC figure.[16] It was an obvious and easily identified mistake, but one that the climate contrarians failed to recognize due to confirmation bias. The draft IPCC figure seemed to indicate what they wanted to believe—that mainstream climate models were overestimating global warming—and thus they didn't question the validity or accuracy of the graph.

As you would expect, scientists reviewing the draft report caught the error, and the IPCC corrected it before publication of the final report. That's the reason the IPCC report has a review process. The incident illustrates why it's not just unethical, but also generally a bad idea to report on draft documents. A draft by definition is subject to change.

This presented a situation where the skepticism of climate contrarians would be tested. Would they acknowledge that they had made a mistake in reporting on the erroneous draft figure, or would they try to defend their mistake and continue to claim that the IPCC climate models were inaccurately overestimating global warming?

The IPCC released the draft final report in question on September 30, 2012. Writing for *The Guardian* newspaper the following day, I documented the change from the draft to draft final figures.[17] In the model-data comparison figure in the draft final document, the IPCC used an appropriate 30-year baseline period (1961–1990). As a result, the new figure showed that global average surface temperature measurements have fallen within the range of projections made by IPCC model simulations, as I have also illustrated in this book.

I also discussed how we know the final figure is accurate while the draft figure was not. The trend in the global surface temperature changes does not depend on the choice of baseline. Thus, if you doubt the accuracy of the visual depiction of global temperature data in a graph, the way to check it is to compare the trends.

From 1990 through 2012, the measured global surface temperature warming trend was about 0.19°C (0.34°F) per decade. In each previous report, the IPCC model simulations produced a range of global surface temperature trend projections. This is because we can't know ahead of time how the model input variables will behave in the future, for example, the internal variability of ocean cycles or the amount of greenhouse gases humans will pump into the atmosphere. Thus, climate scientists run a large number of individual simulations, each of which represents one possible path that the climate could follow. These provide a range of possible future outcomes, and climate scientists expect the real-world climate change to fall somewhere within that range.

In each previous IPCC report, the lower end of the average global surface warming projected by all of those individual climate model simulations was about 0.1°C (0.18°F) per decade. The upper end of the range varied between the different reports, being highest in the 1990 FAR, lowest in the 1995 SAR, and between the two in the 2001 TAR and 2007 AR4. However, in every case, the measured global surface temperature trend has fallen within those lower and upper range model projections. Therefore, regardless of the choice of figure baselines, we know that the observed global surface warming (0.17°C or 0.31°F per decade since 1990) has been consistent with the range of IPCC model projections and has been faster than the lower model simulations (0.1°C or 0.18°F per decade).

An accurately baselined figure should reflect those trends. This tells us that the draft IPCC AR5 figure that seemed to show the temperature data falling below the range of IPCC model simulations was improperly baselined and visually misleading. The IPCC was correct to revise the figure to use a 30-year baseline and reflect the fact that the measured temperature trends have fallen within the range of model-simulated temperature trends.

So how do you think climate contrarians reacted to this news? If you guessed that they reacted with genuine skepticism and open mindedness, acknowledging that they had made a mistake to report on the erroneous draft figure and that the IPCC was correct to revise it, well, then you don't know climate contrarians very well.

The day before my article for *The Guardian*, climate contrarian blogger Steve McIntyre wrote a blog post of his own on the subject.[18] McIntyre built up a significant climate contrarian fan base by criticizing a graph known as "the hockey stick," first created in 1998–1999 by a team of climate scientists led by Michael Mann.[19,20] The graph was among the first reconstructions of surface temperatures across the Northern Hemisphere over the past 1,000–2,000 years and indicated that current temperatures are likely the hottest they've been during that time frame. It became known as the hockey stick because temperatures leading up to the past century were relatively flat, followed by a steep rise over the past 100 years, resembling the shaft and blade of a hockey stick.

Needless to say, climate contrarians don't like the message depicted by the graph, which indicates that the current global warming is unnaturally rapid and heading into uncharted climate territory as compared to the past few millennia. McIntyre teamed up with economist Ross McKitrick in 2005 to publish a paper disputing the validity of the statistical methods used in creating the hockey stick.[21] The validity of the criticisms leveled by McIntyre and McKitrick has been hotly disputed, but it's become a moot point. Subsequent to the work of Mann and colleagues in 1998–1999, there have been dozens of other studies reconstructing Northern Hemisphere and global surface temperatures over the past 1,000–2,000 years. In every case they have found the same general hockey stick shape and concluded that current temperatures are at or near their hottest levels during that time frame.

The most convincing such study was completed by the Past Global Changes (PAGES) network of climate scientists in 2013. The PAGES 2k consortium consists of scientists from nine regional working groups, each of which collects and processes the best paleoclimate (past climate change) data from its respective region. It's a clever approach because it allows the experts in their local data to contribute to a much larger global project.

The PAGES team published a paper in 2013 with contributions from 78 researchers hailing from 60 separate scientific institutions around the world.[22] Their analysis combined records from tree rings, pollen, corals, lake and marine sediments, ice cores, stalagmites and historical documents from 511 locations across seven continental-scale regions to reconstruct global surface temperature changes over the past 2,000 years. Their reconstruction was the most comprehensive to date, and their results very closely matched Mann's original hockey stick. Statistical arguments aside, the result of Mann and his colleagues has been replicated and vindicated.

In any case, McIntyre became a darling of the climate contrarian community for attacking their hockey stick nemesis, as Mann documented in his book *The Hockey Stick and the Climate Wars: Dispatches from the Front Lines*.[23] As a reward for his statistical critiques, contrarians consider McIntyre some sort of mathematical genius. So you would hope that of all climate contrarians, McIntyre would recognize the obvious baselining error in the draft IPCC figure and acknowledge that they were correct to revise it.

Alas, McIntyre's biases instead clouded his judgment. In his blog post about the IPCC figure, not only did McIntyre fail to recognize the baselining adjustment (whereas statistician and climate realist blogger Tamino had immediately recognized the baselining error 10 months earlier, as soon as the draft figure was leaked), but he even implied that the IPCC was trying to mislead the public by shifting the data upward to create the impression that the model projections were accurate. McIntyre's perception was the exact opposite of reality.

McIntyre was not alone in letting his biases cloud his judgment. Climate scientist and contrarian Judith Curry from Georgia Tech, who had frequently falsely proclaimed that the IPCC climate models had been proven inaccurate prior to the release of the AR5 report, similarly refused to acknowledge her mistakes. Rather than analyze the trends as I had done, Curry rejected my explanations and referred to McIntyre's blog post, along with a comment made by McKitrick on that blog, claiming that in the draft figure, "The trend discrepancy was quite visible."[24]

Unlike me, McKitrick hadn't actually analyzed the trends numerically; he simply eyeballed them and decided they looked incompatible. This is a big no-no, as any scientist should know. Scientists don't rely on their senses for a good reason—our senses can easily deceive us. That's what happened with McKitrick. As I showed, the observed global surface warming trends have fallen within the range of IPCC model simulations. That was true in both the draft and final versions of the graph, with the improper baselining of the former being the only significant difference.

Nevertheless, rather than analyzing the data or accepting the trends that I had provided her, contrarian scientist Judith Curry deferred to the gut feelings of McIntyre and McKitrick. Their gut feelings were wrong. This is a very good example of the difference between climate realists and climate contrarians. The former are the true skeptics, basing their conclusions on the full body of scientific evidence. The latter simply have a gut feeling that something must be wrong with the scientific evidence, usually because they don't like the policy implications (that we need to take action to wean ourselves off our fossil fuel addiction). They then cherry-pick the data to support their gut feelings, rejecting any inconvenient data.

It is true that in the relatively short period between 1998 and 2012, global surface temperatures warmed more slowly than most (but not all) climate model simulations projected. Climate contrarians like Judith Curry have focused exclusively on this 15-year period to claim that climate models have failed and to sow doubt in the minds of the public. However, it's also true that between 1992 and 2006, surface temperatures warmed faster than most climate model simulations.

Climate scientists would like to better understand the ocean cycles that are a primary factor in these short-term temperature variations, in order to be able to predict short-term climate changes. However, they make very little difference in terms of long-term climate changes. That's because cycles are cyclical. The faster surface warming from 1992 to 2006 was offset by the slower surface warming from 1998 to 2012. The positive and negative cycles cancel out, which is why climate models did a good job projecting the changes from 1990 to 2012, even though they didn't do very well from 1992 to 2006 or from 1998 to 2012.

It's also interesting to compare the reactions of climate contrarians to the relatively slow 1998–2012 global surface warming to the reactions of mainstream climate scientists (who contrarians would call "alarmists") to the relatively fast 1992–2006 global surface warming. For example, a 2007 paper published by a team led by well-known German climate scientist Stefan Rahmstorf specifically examined the rapid global surface warming from 1990 through 2006.[25] They concluded:

> The global mean surface temperature increase (land and ocean combined) in both the NASA GISS data set and the Hadley Centre/

> Climatic Research Unit data set is 0.33°C for the 16 years since 1990, which is in the upper part of the range projected by the IPCC. . . . *The first candidate reason is intrinsic variability within the climate system.* (emphasis added)

This reveals a stark contrast between the mind-sets of climate contrarians and mainstream climate scientists. Contrarians by their nature try to find something wrong with climate science data and research. Once they find a seeming flaw, rather than try to explain or understand the science behind it, they blow it out of proportion and declare that the whole field of climate science is wrong and that therefore climate change is nothing to worry about.

On the other hand, when mainstream climate scientists see a climate change that doesn't quite match their expectations, they try to understand and explain what's going on. Rahmstorf's team didn't immediately assume that because short-term surface warming was happening faster than expected, that necessarily meant the models weren't sensitive enough to the increased greenhouse effect, or that global warming was going to be worse than expected. Their immediate reaction was that this was probably just short-term noise from natural internal variability. And they were right.

THE POLITICAL IDEOLOGICAL BASIS OF CLIMATE CONTRARIANISM

Nevertheless, the contrarian focus on the short-term global surface warming slowdown has gained a lot of traction in the mainstream media. This has mainly been an issue in politically conservative media outlets, particularly those owned by Rupert Murdoch like Fox News, *The Wall Street Journal*, *The Australian*, and *The Times* in the United Kingdom.

This goes to show that climate contrarianism stems from political rather than scientific grounds. It's not as though journalists at these politically conservative media outlets have some particular insight into climate science. Rather they approach the subject completely backward, beginning with their ideological bias to climate policy solutions. In order to justify their opposition to these policies, contrarians seek out scientific evidence which seems to contradict the science. If there isn't a problem, then we don't need to solve it.

Climate contrarians then cherry-pick that evidence, generally taking it out of context and misrepresenting it, as they have done with the recently slowed global surface warming, and ignore the vast body of evidence that contradicts their predetermined conclusion. In this case, they ignore the fact that short-term global surface temperatures were warming faster than most climate model projections up to 2006 and that when we consider the warming of the entire global climate system (including the oceans, which absorb over 90 percent of that warming), if anything global warming has accelerated.

Unfortunately, the conservative media have placed so much emphasis and run so many stories focusing on the surface warming hiatus that it has begun to seep into the nonconservative media as well. The organization Media Matters for America conducted a study of climate change coverage in the American media during the two months leading up to the publication of the 2014 IPCC AR5 report.[26] The good news from the study was that most American media outlets did a good job covering the story, particularly CNN, which ran over 30 pieces on the IPCC report in a two-month period without succumbing to the temptation of false balance by overrepresenting climate contrarians.

In fact, most large American newspapers and news networks did a good job accurately representing the mainstream scientific consensus on human-caused global warming. The exceptions, not surprisingly, were Fox News and *The Wall Street Journal*, which interviewed and quoted climate contrarians in most of their stories about the IPCC report.

However, the politically conservative media outlets had successfully managed to draw attention to the recent slowed global surface warming. As a result, although it wasn't discussed in most television news stories (with the exception of Fox News, CBS, and a few times by CNN), all print media outlets discussed the hiatus in a significant percentage of their IPCC stories. In addition to *The Wall Street Journal*, the *LA Times*, the *New York Times*, and the *Washington Post* all discussed the surface warming slowdown in at least half of their stories about the IPCC report. This in turn can be attributed in large part to the IPCC itself, which felt enough pressure that it discussed the subject in its Summary for Policymakers. The IPCC said:

> The long-term climate model simulations show a trend in global-mean surface temperature from 1951 to 2012 that agrees with the observed trend (very high confidence). There are, however, differences between simulated and observed trends over periods as short as 10 to 15 years (e.g., 1998 to 2012).[27]

This was a difficult decision for the IPCC, due in large part to the leaked flawed draft figure comparing the global surface warming measurements and climate model projections. Climate contrarians were focusing almost exclusively on the short-term slowed global surface warming, and the IPCC thus felt they needed to address it, which in turn caused more mainstream media news outlets to cover the issue.

This is always a challenge for climate communicators and myth debunkers. When explaining or debunking a myth or piece of climate science misinformation, in the process you also draw more attention to it. If not done carefully, debunking a myth can actually act to reinforce the piece of misinformation in the public mind.

John Cook and Stephan Lewandowsky wrote *The Debunking Handbook* that addresses how to avoid various pitfalls during myth debunking, such as the familiarity backfire effect. This happens when the myth must be repeated in order to debunk it, but this may cause the audience to remember the myth more vividly. The familiarity backfire effect can be avoided by either avoiding repetition of the myth (which can be a challenge when trying to debunk it) or sandwiching it between facts. However, headlines about a global warming hiatus or pause will draw in a large audience, and thus, most mainstream media stories about the subject suffer from the familiarity backfire effect, especially since many people don't read past the headline of a story.

By focusing so heavily on the short-term noisy data, climate contrarians have successfully triggered a domino effect that has drawn much more attention to this minor scientific question. Their focus on the hiatus caused the IPCC to address it, which led to more mainstream media stories about it, whose headlines repeated the myth, which triggered the familiarity backfire effect, which reinforced the myth in the minds of the public.

All of this originated from the politically conservative media working backward to justify their ideological bias to climate solutions. Their nearly exclusive focus on the cherry-picked short-term data has spread to many other news outlets, causing a wider audience to become misinformed in the process. This is the climate misinformation campaign that mainstream climate scientists and communicators are forced to constantly battle against. Spreading misinformation and myths is much easier than successfully debunking those myths with scientific evidence and facts. If debunking is not done carefully, efforts to debunk the myths can actually backfire and reinforce them instead. Another problem is the media fear of being criticized for bias if failing to represent "both sides" in a given story. Sometimes, particularly when it comes to science, two sides of an issue don't have equal validity.

For example, when running stories about the adverse health effects of smoking, it's no longer considered necessary or even good journalism to interview someone representing the side that denies the health effects of smoking. That position has come to be viewed as wrong because it's not supported by scientific evidence. However, climate scientists are just as confident in human-caused global warming as medical doctors are that smoking causes adverse health effects. In both cases, the scientific evidence is overwhelming. Yet many media outlets feel pressure to "balance" stories that discuss mainstream climate science by also interviewing climate contrarians. BBC editor Ehsan Masood said in 2013 that their network would continue to interview climate contrarians for fear of "shutting out dissenting voices."[28] There's certainly nothing wrong with including "dissenting voices" when it comes to debating climate policy and solutions or when discussing an unsettled technical climate science detail. However, when discussing the causes of global warming, including the dissenting voices of climate contrarians, as the BBC has done, is no different than interviewing those who would argue that smoking doesn't have adverse health effects or that HIV doesn't cause AIDS. It's bad journalism that misinforms the audience.

IPCC REPORT ON CLIMATE IMPACTS AND ADAPTATION

In addition to the IPCC Physical Basis report that summarizes our understanding of how the climate is changing, the IPCC publishes a second report on climate impacts and adaptation and a third report on climate change mitigation. The second report, on climate impacts adaptation, was published in late March 2014. The report painted a rather bleak picture of our future climate.

For example, the report discussed the risks associated with food insecurity due to more intense droughts, floods, and heat waves in a warmer world, especially for poorer countries. This contradicts the claims of many climate contrarians who have tried to claim that rising carbon dioxide levels are good for crops. It's true that carbon dioxide acts as "plant food" in a greenhouse setting where we can control all other factors (temperature, humidity, etc.); the situation outside in the global climate is not so simple. There has been some global "greening" so far from the rise in carbon dioxide, but other factors also influence plant growth, and those factors are playing an increasingly large role as the planet continues to warm.

For example, the increased frequency and/or intensity of heat waves and floods and droughts that result from human-caused global warming tend to be detrimental to plant growth. Up to this point the carbon dioxide plant fertilization effect has won out, but research indicates that this trend may already be reversing. Technological improvements have also allowed us to dramatically increase crop yields, for example, through the use of nitrogen-based fertilizers. However, there's a limit to how much food we can produce from a given amount of available agricultural land even with improving technology, and now farmers also have to battle the aforementioned effects of climate change like worse heat waves and more intense droughts.

The IPCC report also discussed risks associated with water insecurity, due, for example, to shrinking of glaciers that act as key water resources for various regions around the world and through changing precipitation patterns. As a result of these types of changes, the IPCC anticipates that violent conflicts like civil wars will become more common. Syria is a prime example, with a severe drought in the region between 2006 and 2011 amplified by human-caused global warming, leading to a collapse of its farms and livestock.[29] The Syrian government failed to assist its impacted farmers, creating widespread civil unrest, leading to a civil war.

The IPCC also projects that the number of people exposed to river floods will increase as the planet warms over the remainder of the century. Sea level rise will cause submergence, flooding, and erosion of coastal regions and low-lying areas. And ocean acidification poses significant risk for marine ecosystems, coral reefs in particular. In fact, the general risk of species extinctions rises as the planet warms. More climate change means that suitable climates for species shift. The faster these climate zones shift, the more species will be unable to track and adapt to those changes. The latest IPCC report said:

> Many species will be unable to track suitable climates under mid- and high-range rates of climate change (i.e., RCP4.5, 6.0, and 8.5) during the 21st century (medium confidence). Lower rates of change (i.e., RCP2.6) will pose fewer problems.[30]

The IPCC report also estimated that global surface warming of approximately 2°C (3.6°F) above current temperatures may lead to global income losses of 0.2 percent to 2.0 percent. However,

> Losses are more likely than not to be greater, rather than smaller, than this range . . . few quantitative estimates have been completed for additional warming around 3°C or above.[31]

Even in the IPCC's most aggressive greenhouse gas emissions reductions scenario, we limit global warming only to around 1°C (1.8°F) above current temperatures. In a business-as-usual (BAU) scenario, temperatures warm about another 4°C (7.2°F)—yet we have difficultly estimating the costs of warming exceeding another 2 to 3°C (3.6 to 5.4°F). In other words, failing to curb human-caused global warming poses major risks to the global economy.

Nevertheless, there will be a certain amount of climate change that we won't be able to avoid, and the IPCC report noted that adaptation to those changes is also critically important. The problem is that many journalists writing about the IPCC report were either looking for a "balanced" approach or trying to downplay the risks posed by human-caused climate change. As a result, many stories claimed that the IPCC was suggesting we should expend less effort trying to mitigate global warming and instead simply try to adapt to it. These stories were examples of absolutely terrible journalism—just two weeks after publishing its report on climate impacts and adaptation, the IPCC published its report on climate mitigation.

This wasn't a surprise—the IPCC had publicized this report schedule and has always published separate reports on impacts/adaptation and mitigation/response. For example, in 1990, the IPCC FAR published reports on *The Scientific Assessment of Climate Change, Impacts Assessment*, and *Response Strategies*. Since the TAR in 2001, the second report has been on *Impacts, Adaptation and Vulnerability*, and the third report has been on *Mitigation of Climate Change*. For journalists to claim that the publication of the second report in 2014 indicated the IPCC was now shifting its recommendations away from mitigation toward adaptation was grossly incompetent, as was illustrated a mere two weeks later when the IPCC report on climate mitigation was published.

Many other media stories about the IPCC impacts and adaptation report focused on Richard Tol, who the IPCC had perhaps unwisely made a lead author on one of its chapters. While Tol is a qualified climate economist, his research is also an outlier, as Tol has published the only papers finding that modest global warming could result in a significant net benefit to the economy (based on large part on the assumption that crop yields will continue to increase). Tol had also previously made strong critical comments about the IPCC in general and is an advisor to a UK anti-climate policy advocacy group, the aforementioned Global Warming Policy Foundation.

Tol is also known as a disruptive influence, as he showed when he attacked our consensus paper. Tufts University economist Frank Ackerman also became one of Tol's targets after he and his colleague

Charles Munitz dared publish a paper detailing some flaws in Tol's climate economics model called FUND.[32] Ackerman and Munitz found two potential flaws in Tol's model, which they argued:

> The defects we identified both tend to exaggerate the benefits of climate change for agriculture.

Tol's FUND is the only major climate economics model concluding that a degree or two of global warming could result in a significant net benefit to the global economy, in large part because it concludes that some climate change could result in higher agricultural productivity. If Ackerman and Munitz are right about this flaw in his model, it could explain why its results are an outlier.

Tol's response to this critique was aggressive. First, he asked the journal to issue a retraction or correction of the critique. Instead, the journal editor published a letter on the controversy listing the points that Tol alleged Ackerman and Munitz got wrong, the points that Ackerman and Munitz agreed to change as a result, and the points where the two sides could not resolve their disagreements.[33]

Unfortunately, Tol didn't stop there. He also wrote to Ackerman's employer and publishers, accusing him of libel for writing his technical critique of FUND. Tol circulated the journal editor's letter on the controversy, alleging that it proved the article was libelous, which it did not. Ackerman and Muniz's paper had first appeared as a Stockholm Environment Institute (SEI) working paper. Tol sent numerous e-mails and letters to SEI, to the vice-chancellor of Stockholm University, to the Swedish Royal Academy of Arts and Sciences, to the Swedish Minister of Environment, and to the Swedish Minister of Education, demanding that SEI withdraw or rewrite the working paper, publish his reply, and apologize to him. SEI executive director Johan Kuylenstierna and former executive director Johan Rockström ultimately wrote a letter on the matter, stating:[34]

> Professor Tol's repeated, groundless attacks on this article, and on SEI for its association with the article, have violated the norms of civility and scholarly debate. We urge Professor Tol to stop attacking the motivations and reputations of others, and to return to the academic community that accepts disagreement and engages in substantive debate.

Given this behavior and the fact that Tol's research is an outlier in minimizing the economic costs of climate change, many were puzzled when the IPCC selected him as a lead author on one of its chapters. As might have been expected, Tol was a disruptive influence, removing his name from the report's Summary for Policymakers and refusing to participate in its drafting process. This move became the focus of many media stories, which claimed that several IPCC authors were dissatisfied with the report, even though Tol was the only author identified as being critical of the final product. Controversy sells in the media, and many journalists decided to focus on Tol rather than focusing on the IPCC report itself. The IPCC probably regretted making Tol a lead author on one of its chapters.

In any case, the good news is that the IPCC report concluded that many climate risks can be reduced if we act to slow global warming and thus avoid the worst climate change scenarios. For example, the IPCC stated with high confidence that risks associated with reduced agricultural yields, water scarcity, inundation of coastal infrastructure from sea level rise, and adverse impacts from heat waves, floods, and droughts can be reduced by cutting human greenhouse gas emissions.

In the end it all boils down to risk management. The stronger our efforts to reduce greenhouse gas emissions, the lower the risk of extreme climate impacts. The higher our emissions, the bigger climate changes we'll face, which also means more expensive adaptation, more species extinctions, more food and water insecurities, more income losses, more conflicts, and so forth. As Chris Field, cochair of the IPCC report on climate impacts and adaptation, noted,

> With high levels of warming that result from continued growth in greenhouse gas emissions, risks will be challenging to manage, and even serious, sustained investments in adaptation will face limits.[35]

IPCC REPORT ON MITIGATION

The IPCC published its report on mitigation just a few weeks after the report on impacts and adaptation, as scheduled. The report actually included some very good news. Although we're running out of time, if we act now, avoiding the most dangerous levels of global warming can be done very cheaply. Specifically, the report concluded:

> Mitigation scenarios that reach atmospheric concentrations of about 450ppm CO_2eq by 2100 entail losses in global consumption—not including benefits of reduced climate change as well as cobenefits and adverse side-effects of mitigation . . . [that] correspond to

> an annualized reduction of consumption growth by 0.04 to 0.14 (median: 0.06) percentage points over the century relative to annualized consumption growth in the baseline that is between 1.6% and 3% per year.[36]

In other words, the global economy grows at a rate of around 2.3 percent per year, and we can prevent global warming from surpassing the 2°C (3.6°F) "danger limit" while the global economy continues to grow at a rate of about 2.24 percent per year. In fact, the economy may even continue to grow faster than that, because the IPCC isn't accounting for the incidental economic benefits of reducing greenhouse gas emissions. For example, the cleaner air and water and associated health benefits that come with transitioning away from dirty high-carbon energy sources save money. Low carbon energy sources also tend to generate more jobs than fossil fuel energy sources.

Unfortunately, many media outlets tried to compare the economic figures in the IPCC adaptation and mitigation reports and incorrectly concluded that it's cheaper to try and adapt to climate change than to prevent it. Several newspapers printed comments from Bjorn Lomborg, for example, who wrongly claimed in an interview with Rupert Murdoch's *The Australian*,[37]

> If we don't do anything, the damages caused by climate change will cost less than 2 per cent of GDP in about 2070. Yet the cost of doing something will likely be higher than 6 per cent of GDP, according to the IPCC report.

In reality, that's not at all what the IPCC report said. The problem is that the second IPCC report estimated the costs of adaptation in terms of annual income losses, whereas the third IPCC report estimated the costs of preventing global warming by reducing greenhouse gas emissions in terms of slowed economic growth. These figures aren't directly comparable—they're apples and oranges.

To sort these numbers out, I spoke with Cambridge University climate economist Chris Hope, who told me that if the goal is to figure out the economically optimal amount of global warming mitigation, the IPCC reports "don't take us far down this road."[38] To do this comparison properly, the benefits of reduced climate damages and the costs of reduced greenhouse gas emissions need to be compared in terms of "net present value." That's the sort of estimate Integrated Assessment Models like Hope's PAGE were set up to make.

According to Hope's model, the economically optimal peak atmospheric carbon dioxide concentration is around 450 to 500 ppm, with a peak global surface warming of about 2.5 to 3°C (4.5 to 54°F) above preindustrial temperatures (about 1.5 to 2°C or 2.7 to 3.6°F warmer than present). In his book *The Climate Casino*, Yale economist William Nordhaus notes that he has arrived at a similar conclusion in his modeling research.

To limit global warming to that level would require major efforts to reduce greenhouse gas emissions, but as the IPCC report on mitigation noted, that would only slow the global economic growth rate by a little bit. It would also only slow economic growth as compared to a fictional world in which global warming and climate change have zero impact on the economy. In reality, we should be comparing the costs of climate policies to the costs of letting global warming continue to wreak havoc on the climate. According to the economic models run by Hope and Nordhaus, the slowed economic growth rate from reducing carbon pollution would be more than offset by the savings from avoiding climate damages above 2.5 to 3°C global warming.

Although the IPCC didn't make this comparison, these economic modeling results are consistent with its reports. The IPCC report on adaptation was only able to estimate the costs of climate damages for an additional 2°C of global warming and noted that beyond that point, the costs accelerate to a point where they become very difficult to estimate. Nordhaus has similarly noted:[39]

> In reality, estimates of damage functions are virtually non-existent for temperature increases above 3°C.

Without the modeling tools used by economists like Hope and Nordhaus, the figures in the two IPCC reports can't properly be put into an apples to apples comparison. That was Lomborg's first problem.

Lomborg's second problem was in also cherry-picking the year 2070 to make his economic comparison between the costs of global warming adaptation and mitigation. Why 2070? By that point, in a BAU scenario the planet probably won't have warmed much more than 2°C compared to current temperatures. The problem with this cherry-pick is that the world won't end in 2070 (hopefully!); in fact, most of today's children will still be alive in 2070. If we continue on that BAU path, global warming will continue to accelerate after 2070, past the point where economists can't even accurately estimate its accelerating costs.

Lomborg's cherry-pick completely disregards the welfare of future generations.

The bottom line is that economists can't even accurately estimate how much climate damages will cost if we fail to take serious steps to slow global warming. On the other hand, taking those steps can have a negligible impact on global economic growth. It's important to understand that our choices aren't to either reduce carbon emissions or do nothing. Our options are to either reduce carbon emissions or continue with BAU emissions that will cause accelerating climate change and damage costs beyond what we can accurately estimate. From an economic perspective, and from a risk management perspective, this should be a no-brainer. As economist Paul Krugman put it,[40]

> So is the climate threat solved? Well, it should be. The science is solid; the technology is there; the economics look far more favorable than anyone expected. All that stands in the way of saving the planet is a combination of ignorance, prejudice and vested interests. What could go wrong?

It's an interesting contrast that all the economics experts agree that tackling global warming can be relatively cheap, whereas continuing on a BAU path will cause climate damages so large that they can't even be properly estimated. Meanwhile the media interview nonexperts like Bjorn Lomborg, who assure us that the opposite is true and we'd be better off just letting global warming continue unabated.

This is another perfect example of the problem with false balance in the media. Journalists tend to want to appear "balanced" in their reporting and thus look for seemingly credible voices to present "the other side." Lomborg is an author who's well known for presenting this rosy outlook, always downplaying the dangers of climate change for the benefit of journalists seeking false balance. However, he's not an economist, and it seems as though the only reason he has been able to become a seemingly credible voice is due to the attention journalists have given him.

A sort of symbiotic relationship between journalists and contrarians has formed. In the age of Fox News, journalists are considered credible if they're "fair and balanced." Factual accuracy in reporting seems to have become a secondary consideration behind the desire for balance. Lomborg and several other nonexperts like him provide the media with that "balanced" perspective, and the journalists return the favor by printing their comments in major news outlets, thereby artificially boosting their perceived credibility.

While this approach might make sense for an opinion-based subject like politics, it doesn't work for fact- and evidence-based subjects like science. When a media outlet runs a story that discusses evolution or the health effects of smoking, we don't expect them to also include comments from creationists or people who deny that smoking causes cancer. On those subjects we accept that the scientific evidence is conclusive and there's no reason to include the opinions of those who deny it.

Yet many journalists continue to view climate science through a political lens. Perhaps this is due to the partisan split on the issue—for ideological reasons, those who are politically conservative-minded are much more likely to reject the realities of human-caused climate change than those who are politically liberal-minded. However, science doesn't have a political bias, and that includes climate science. The expert consensus on human-caused global warming is a result of the overwhelming scientific evidence supporting it and has nothing whatsoever to do with politics. That's just how science works.

Likewise, economists support tackling climate change by putting a price on carbon emissions, because that's what the economics research indicates is the most economically beneficial approach. To make the contrary argument, journalists are forced to turn to nonexperts like Bjorn Lomborg who do the calculations wrong. The numbers simply don't support continuing on a BAU path. Economists can't even accurately estimate the costs of global warming damages above 3°C (5.4°F), and limiting global warming to that level will require major international efforts to reduce greenhouse gas emissions.

The good news is that those efforts can be made at with very little impact to global economic growth. Apparently that news isn't controversial enough for many journalists, who instead turn to nonexperts like Bjorn Lomborg who will give them the material for the story they want to write, even though that story isn't accurate.

PRESENT DAY

It's interesting and informative to compare all the global temperature predictions we've examined to this point, to see which have been the most accurate. McLean's prediction was by far the worst, predicting an enormous amount of cooling in 2013 that obviously didn't happen. Not surprisingly, Akasofu, who predicted ongoing global warming, has been closest to reality among the climate "skeptics." However, every single contrarian prediction has already underestimated global warming, and almost every mainstream climate science prediction has been more accurate (with Kellogg in 1979 the one exception).

Considering that they represent a summary of the best climate science done to date, it's also not surprising that the IPCC reports have generally produced the most accurate predictions thus far. In addition, the models with a climate sensitivity of close to 3°C (5.4°F) for a doubling of atmospheric carbon dioxide have generally been the most accurate. For example, Hansen's 1988 model, with its sensitivity of 4.2°C (7.6°F), predicted more warming than has been observed.

Despite their consistent inaccuracy, the climate contrarians never seem to lose credibility. They're frequently referenced by the media in an attempt to be "fair and balanced," and as a result, many people think there is significant debate among climate scientists regarding whether the planet is warming, what's causing that warming, and whether the warming will continue. In reality, it's only a few fringe contrarians who dispute the scientific consensus that humans are causing dangerous global warming, and these contrarians have a long history of making inaccurate claims and predictions. We once again return to the sage words of George Santayana:

> Those who cannot remember the past are condemned to repeat it.[41]

If we continue to listen to these contrarians despite their history of wrong assertions and predictions, we will fail to sufficiently reduce our greenhouse gas emissions, and the planet will continue to warm at a dangerous rate.

ATTEMPTS AT CREDIBILITY THROUGH PAL REVIEW

A common refrain among climate contrarians is that the 97 percent expert consensus in the peer-reviewed scientific literature on human-caused global warming is a result of "pal review" or "gatekeeping." Basically, they argue that mainstream climate scientists run all of the big climate science journals, and they'll publish papers from their "pals" and reject papers from climate contrarians.

This is really a silly conspiracy theory. There are examples throughout this book of climate contrarians getting their (scientifically poor) papers published in scientific journals. Contrarians get government research grants, they do scientific research at government academic institutions, and then they publish their results in peer-reviewed journals. Mainstream climate scientists do the same thing.

What contrarians call "gatekeepers" are journal editors who try to make sure that papers submitted to their journals go through a rigorous review by experts in the applicable scientific field. Some papers aren't good enough to be published in quality journals—not just contrarian papers, but many papers by mainstream climate scientists are rejected by journals as well. That's just how the peer-review process works. The top journals reject two-thirds or more of the scientific papers they receive as submissions.

There have been examples of pal review, where a scientist or group of scientists find a friendly journal editor who will send their papers out to friendly reviewers and thus make it easy for their material to get published in that journal. However, in most cases, this pal review abuse of the peer-review system comes not from mainstream climate scientists, but instead from climate contrarians. The journal *Climate Research* was a prime example.

During most of the period from 1997 to 2003, *Climate Research* didn't have an editor in chief. Instead, scientists who wanted to publish a paper in this journal sent their manuscripts to an associate editor of their choice. My colleague John Mashey researched the papers published by *Climate Research* during that six-year period.[42] Mashey found that one particular associate editor, Chris de Freitas, published 14 separate papers from a select group of 14 climate contrarians during that time. People who are familiar with the climate "debate" will recognize the names of many of these contrarian pals.

- Patrick Michaels
- Willie Soon
- Sallie Baliunas
- John Christy
- David Douglass
- Ross McKitrick
- Robert Balling
- Robert Davis (both *Climate Research* author and editor)
- Vincent Gray
- Sherwood Idso
- PJ "Chip" Knappenberger
- Eric Posmentier
- Arthur Robinson
- Gerd-Rainer Weber

Patrick Michaels, who now works for the Cato Institute conservative advocacy group, was an author on 7 of the 14 pal-reviewed papers. These accounted for half of his overall peer-reviewed publications from 1997 to 2003. During this six-year period, 14 of the 24 papers accepted by Chris de Freitas for publication in *Climate Research* came from this group of 14 climate contrarians. Basically, the contrarians had found a friendly journal editor who would make publication of their papers (which frequently disputed human-caused global warming) much easier than submitting their papers to journals with more stringent peer-review processes.

It all came to a head when *Climate Research* published a fundamentally flawed paper by Willie Soon and Sallie Baliunas in 2003.[43] Their paper reviewed 200 other research papers dealing with climate changes over the past 1,000 years and concluded that the global warming and climate change we've experienced over the past century is nothing remarkable. This conclusion was in direct contradiction with all other research that reconstructed Northern Hemisphere and global surface temperature changes over the past millennium, all of which indicated that recent climate changes have been larger than at any time during that period and resemble a hockey stick shape.

When the Soon and Baliunas paper was published, a number of the scientists whose research was used in their paper complained that their work had been misrepresented. The paper made several unsupportable assumptions, for example, equating dryness with hotness, and was subsequently roundly refuted by an article in the American Geophysical Union journal *Eos* written by a number of prominent climate scientists.[44] Quite simply, the paper never should have been published without first undergoing major corrections to address its fundamental flaws.

However, the conclusions of the paper were very convenient for climate contrarians. At the time it was published, Congress was considering legislation to address climate change. Willie Soon was invited by Senator James Inhofe to testify before Congress, and the Soon and Baliunas paper was used by congressional Republicans to justify opposition to climate legislation. At the time, Senator Inhofe called human-caused global warming "the greatest hoax ever perpetrated on the American people."[45] Obviously, the climate bill was voted down by Congress.

At this time, Hans von Storch had recently been appointed editor in chief at *Climate Research* after working at the journal for 10 years. He pushed for the journal to at minimum require Soon and Baliunas to revise their paper, and to revise the flawed peer-review process, but to no avail. As a result, along with five other editors of *Climate Research*, von Storch resigned from the journal, explaining the reason behind his resignation.

> The reason was that I as newly appointed Editor-in-Chief wanted to make public that the publication of the Soon & Baliunas article was an error, and that the review process at Climate Research would be changed in order to avoid similar failures. The review process had utterly failed; important questions have not been asked. . . . It was not the first time that the process had failed, but it was the most severe case. . . . I withdrew also as editor because I learned during the conflict that [Climate Research] editors used different scales for judging the validity of an article. Some editors considered the problem of the Soon & Baliunas paper as merely a problem of "opinion", while it was really a problem of severe methodological flaws. Thus, I decided that I had to disconnect from that journal, which I had served proudly for about 10 years.[46]

The resignation of such a large fraction of the journal's editorial board was a major scandal in the climate science community. The reputation of *Climate Research* was irreparably damaged (and rightly so), and after 2003 the journal published only a few more papers from the 14 climate contrarian pals.

In 2009 during what came to be known as "Climategate," thousands of private e-mails were stolen from climate scientists at the University of East Anglia in the United Kingdom. Among these e-mails were complaints among climate scientists that something had to be done about the pal review problem at *Climate Research* because it was allowing fundamentally scientifically flawed contrarian papers to be published in a peer-reviewed journal.

By 2009, most people had forgotten about or had never heard of the *Climate Research* pal review scandal. The climate scientists whose e-mails were stolen were accused of gatekeeping and trying to prevent contrarian papers from being published. In reality, these scientists were simply complaining about a problem that would soon create a major scandal at the journal, but by taking their e-mails out of context, climate contrarians successfully turned the tables, making themselves seem like the victims and eroding public trust in the mainstream climate science community.

Nine separate Climategate investigations were conducted and the climate scientists involved were all entirely vindicated, but the damage to the public perception of climate science was done. Climate contrarians had managed to turn their own pal review scandal into an erosion of public trust of the climate scientists who were trying to expose the scandal at the time. Climate contrarians may not be very good at science, but they excel at public relations and misinforming the public.

After the pal review well at *Climate Research* dried up, the next strategy for climate contrarians trying to publish flawed research was to find journals that aren't specific to climate science. Physics journals are quite often the target of climate contrarian publications. The three studies for which I've published papers critiquing their mistakes have all been published in physics journals. The strategy makes sense because physics journals often have good reputations, but the editors of those journals may have difficulty identifying climate science experts to give the submitted papers a proper expert peer-review. If the papers are instead sent out to physicists to review, those physicists will be more likely to miss a key fundamental climate science flaw that a climate scientist reviewer would catch.

In one example, contrarian climate scientists Roy Spencer and Danny Braswell published a paper in 2011 in the journal *Remote Sensing*,[47] which is a journal that focuses on the science and application of remote sensing technology. Their paper dealt with observations made by satellites and was therefore applicable to this journal, but the journal editors nevertheless may not have known any climate scientists who could give the paper a proper expert review.

Climate scientist Kevin Trenberth described *Remote Sensing* as "a fine journal for geographers, but it does not deal with atmospheric and climate science."[48] In fact, one *Remote Sensing* editor, Wolfgang Wagner wrote that "the editorial team unintentionally selected three reviewers who probably share some climate sceptic notions of the authors."[49] Wagner resigned as editor of the journal due to its publication of this fundamentally flawed paper.

The Spencer and Braswell paper used an overly simple climate model and concluded that the climate sensitivity to the increased greenhouse effect is very low. However, the paper had some severe shortcomings, as described by Kevin Trenberth:[50]

> The basic material in the paper has very basic shortcomings because no statistical significance of results, error bars, or uncertainties are given either in the figures or discussed in the text. Moreover the description of the methods in the paper is not sufficient to be able to replicate the results.

Climate researchers Kevin Trenberth, John Fasullo, and John Abraham published a paper showing that Spencer and Braswell had incorrectly treated factors that are a *response* to warming (feedbacks) associated with the El Niño Southern Oscillation as factors that *cause* global warming (forcings).[51] They concluded that the Spencer and Braswell paper was fundamentally flawed in several ways, including using a far-too-simple climate model, not including sufficient details to allow other scientists to replicate their methods and results, and failing to investigate whether changes associated with the El Niño Southern Oscillation were forcings or feedbacks.

Another journal recently targeted by climate contrarians is a rather obscure Korean journal, the *Asia-Pacific Journal of Atmospheric Science*. In 2009, Richard Lindzen and his colleague Yong-Sang Choi published a paper, like Spencer and Braswell's, claiming the climate sensitivity to the increased greenhouse effect is very low.[52] The fundamental flaws in that study were quickly revealed in a paper published by Trenberth and Fasullo along with their colleagues O'Dell and Wong,[53] and by several other papers.

Lindzen himself has even gone as far as to admit the paper contained "some stupid mistakes. . . . It was just embarrassing."[54] The main problem in their analysis involved estimating global climate sensitivity using data from just the tropics. Trenberth and colleagues also showed that the low sensitivity result worked only if the start and end points of the data analysis were carefully chosen (cherry-picked). Trenberth replicated the analysis using different start and end points and got a completely different result.

Lindzen and Choi subsequently put together a new version of the same analysis, attempting to address the critiques of their previous paper. They submitted their paper to a respected climate science journal, but it was rejected. They then submitted it to the *Proceedings of the National Academy of Science* (*PNAS*), which allows members (Lindzen is a National Academy of Science member) to suggest their own reviewers. Lindzen suggested a colleague with whom he had previously published research, and physicist William Happer, who is a physics expert but who has zero climate science expertise or publications and is also an outspoken climate contrarian. As a result, the *PNAS* editors also sent the paper out to two expert climate reviewers.

In the end, all four of the *PNAS* reviewers the—even those chosen by Lindzen himself—unanimously agreed that the journal should not publish the study because it had not adequately addressed the critiques of their 2009 paper. For example, Lindzen and Choi didn't address the point illustrated by Trenberth and colleagues that choosing different start and end points completely changed their result.

After this second rejection, Lindzen and Choi continued shopping their paper around, finally finding a journal willing to publish it in the *Asia-Pacific Journal of Atmospheric Science*. A couple of years later, Spencer and Braswell published another paper not dissimilar from their 2011 *Remote Sensing* paper that was heavily criticized and resulted in the resignation of editor Wolfgang Wagner. Spencer and Braswell again used an excessively simple climate model and concluded that climate sensitivity is low. This time their paper was published in the same *Asia-Pacific Journal of Atmospheric Science* as Lindzen and Choi's flawed study.[55]

Lindzen and Spencer share many other similarities. Like Richard Lindzen, Roy Spencer has made his ideological biases quite clear. For example, he admitted in 2011 "I love FoxNews,"[56] which given the network's frequent antiscience coverage is a surprising statement for any scientist to make. However, Spencer has also published a free-market economics book called *Fundanomics: The Free Market, Simplified*, and on his blog, he has commented:[57]

> I view my job a little like a legislator, supported by the taxpayer, to protect the interests of the taxpayer and to minimize the role of government.

In February 2014, Spencer also launched into a rant on his blog, entitled *Time to Push Back against the Global Warming Nazis*.[58] In that extremely offensive post, Spencer wrote:

> Like the Nazis, they advocate the supreme authority of the state (fascism), which in turn supports their scientific research to support their cause (in the 1930s, it was superiority of the white race).

Clearly Roy Spencer has really made little effort to hide his intense ideological biases. Isn't it interesting how the few contrarian climate scientists so often seem to share the same antigovernment ideologies? It's quite the coincidence.

As exemplified by the papers recently published by Spencer and Lindzen, since they lost their pal review *Climate Research* journal, the strategy of choice for climate contrarians with flawed papers has involved submitting those papers to non-climate and often obscure journals. That way they can avoid rigorous expert peer-review, but still assert that their papers have been published in peer-reviewed journals. However, in 2013–2014, a group of climate contrarians came up with a new clever but short-lived publication strategy.

A few climate contrarians submitted a new journal idea to the respected publisher Copernicus Publications, suggesting creating a new journal that they would call *Pattern Recognition in Physics*. Two individuals proposed becoming editors in chief of this new journal. One was Nils-Axel Mörner, who is best known for denying that sea level is rising and for claiming to be an expert in dowsing (also known as divining, which involves finding objects like groundwater and gemstones underground with a Y- or L-shaped stick).[59] The other was Sid Ali Ouadfeul, who works for the Algerian Petroleum Institute and has been guilty of self-plagiarism in his papers.[60] Neither has any expertise in the field of pattern recognition.

The publishers at Copernicus were worried that Mörner and Ouadfeul would turn their new journal into a global warming–denying paper factory as had happened with *Climate Research*. Mörner and Ouadfeul assured the publishers that their journal would "publish articles about patterns recognized in the full spectrum of physical disciplines rather than to focus on climate-research-related topics."[61] The publishers at Copernicus thus decided to give them a chance with this new journal. It wasn't long before they came to regret that decision.

In only its second edition, the new journal published a "special edition" that was full of papers by authors who are also climate contrarians. Several of the papers disputed human-caused global warming and even the very existence of global warming. The journal engaged in what Copernicus described as

> select[ing] the referees on a nepotistic basis, which we regard as malpractice in scientific publishing and not in accordance with our publication ethics we expect to be followed by the editors.[62]

In essence, the authors reviewed each other's papers. Many of the papers engaged in the same type of "climastrology" as Loehle and Scafetta, described earlier in this book. In fact, Scafetta authored and probably reviewed several of the papers in the *Pattern Recognition in Physics* special edition. It seems that the journal authors and editors interpreted "pattern recognition in physics" to mean finding random astrological cycles in the solar system that match climate cycles on Earth and then claiming, without any supporting physical evidence, that the two are causally related.

The publishers at Copernicus did not appreciate the editors of *Pattern Recognition in Physics* reneging on their promise to include a wide spectrum of physical disciplines and instead turning the journal into a climate contrarian paper mill. Immediately after the special edition was published, Copernicus ceased publication of the new journal. The climate contrarians had come up with a clever idea to start their own journal under a respected publisher, but they had blown it by making their strategy of publishing climate contrarian papers too obvious to miss. Sid Ali Ouadfeul and many of the contributors to *Pattern Recognition in Physics* have tried to continue publication of the journal on their own, but after Copernicus rightfully dropped them, the journal has no remaining credibility. No paper published in *Pattern Recognition in Physics* from here on out will be considered remotely credible.

These various contrarian strategies show that they recognize the importance of peer-reviewed publications in establishing credibility. Although they represent less than 3 percent of peer-reviewed climate publications, when they do publish a paper, contrarians receive a disproportionate amount of media coverage. This is because there are thousands of new papers every year concluding that human-caused global warming is real and is a problem. It's old news.

On the contrary, controversy sells. When a new paper claims to overturn the established science, that makes for a juicy story, and hence, the few contrarian papers tend to receive undeserved media attention. In their effort to publish a popular story, journalists will often fail to look into the history of the publishing authors and hence will often fail to recognize that they have repeatedly been proven wrong like Lindzen, Spencer, and Scafetta. It's a problem that the reputations of climate contrarians aren't dependent upon the accuracy of their science.

The good news is that some media outlets have begun reversing course, moving away from a false balance approach and toward holding climate contrarians accountable for their lack of scientific credibility. For example, in October 2013, *Los Angeles Times* editor Paul Thornton wrote that letters to the editor "that have an untrue basis (for example, ones that say there's no sign humans have caused climate change) do not get printed."[63] This didn't go over well with climate contrarians, so a few days later, Thornton clarified his position.[64]

> As for letters on climate change, we do get plenty from those who deny global warming. And to say they "deny" it might be an understatement: Many say climate change is a hoax, a scheme by liberals to curtail personal freedom . . . scientists have provided ample evidence that human activity is indeed linked to climate change. Just last month, the Intergovernmental Panel on Climate Change—a body made up of the world's top climate scientists—said it was 95%

> certain that we fossil-fuel-burning humans are driving global warming. The debate right now isn't whether this evidence exists (clearly, it does) but what this evidence means for us.
>
> Simply put, I do my best to keep errors of fact off the letters page; when one does run, a correction is published. Saying "there's no sign humans have caused climate change" is not stating an opinion, it's asserting a factual inaccuracy.

A few weeks later, the letters editors of the *Sydney Morning Herald* issued a similar statement.[65]

> Climate change deniers or sceptics are free to express opinions and political views on our page but not to misrepresent facts. This applies to all our contributors on any subject. On that basis, a letter that says, "there is no sign humans have caused climate change" would not make the grade for our page.

These were significant decisions, because the "Letters to the Editor" is one of the most frequently read sections of a newspaper. For that very reason, some climate contrarian groups like the International Climate Science Coalition (of which John McLean is a member, as previously discussed) have specifically targeted the Letters to the Editor section of major newspapers with letters that dispute the expert consensus on human-caused global warming.[66] Thus, it's a good sign that several major newspapers are wising up to this strategy and refusing to run factually inaccurate letters from climate contrarians who deny the scientific reality of human-caused global warming.

Not surprisingly, these decisions by the *LA Times* and the *Sydney Morning Herald* drew heavy criticism from climate contrarian groups and conservative media outlets, who accused them of censorship and oppressing free speech. These accusations are of course baseless; newspapers are free to choose what to print and what not to print on their pages. They're under no obligation to print material that denies human-caused global warming, or evolution, or that smoking causes lung cancer, or that the Earth revolves around the sun. In fact, the editors of these two newspapers are correct that printing this sort of factually inaccurate submission does their readers a disservice and that refusing to print misinformation is part of the job of a good letters editor. People are free to deny whatever scientific realities they like, shouting their denial from the rooftops in a public space. That's freedom of speech. But newspapers are under no obligation to print science denial on their pages.

Unfortunately, that hasn't stopped many (mostly politically conservative) media outlets from airing the scientifically incorrect and unsupported opinions of climate contrarians like Richard Lindzen, Roy Spencer, and John McLean. The fact that they each have long histories of being wrong on climate science doesn't seem to damage their credibility in the eyes of certain media outlets that want to spread doubt about human-caused global warming. All it takes is a job title as "climate scientist" and a contrarian position on global warming, and politically conservative news outlets like Fox News and *The Wall Street Journal* will scramble to write a story about the maverick scientist who's bucking the consensus and assures us all that global warming is nothing to worry about. The scientific accuracy of that contrarian's arguments isn't considered relevant, as long as he or she feeds those biased media outlets the story they want to tell. While conservative media outlets are free to print factual inaccuracies and myths from climate contrarians, they misinform their readers by doing so.

The good news is that there are many good media outlets like the *LA Times*, the *Sydney Morning Herald*, and *The Guardian* that will take a scientist's accuracy and credibility into account and strive not to publish factually inaccurate letters, editorials, or stories. The bad news is that there are plenty of poor, generally politically conservative media outlets that don't maintain that same high standard. The other problem is that a significant number of people (who are generally also politically conservative) get most of their information from these politically conservative media outlets. Thus, the low standard of scientific and factual accuracy in these media outlets when it comes to the subject of climate change is causing a significant fraction of the population to become misinformed on this critical subject.

7

What Does the Future Hold?

Since the IPCC projections have thus far been the most accurate, it's worthwhile to examine their projections about we have in store for the future. According to the latest IPCC report, if we continue on a business-as-usual (BAU) path relying on fossil fuels, we could see 5°C (9°F) global surface warming above preindustrial levels by 2100. The good news is that if we take serious steps to reduce carbon pollution, we can limit global warming to the "danger limit" target of no more than 2°C (3.6°F).

Determining at exactly what point global warming will become very harmful or catastrophic is a difficult task. However, the IPCC has summarized some of the impacts that climate scientists expect to see once the planet warms to certain levels above preindustrial temperatures. Most of the worst consequences are expected to strike once the planet has warmed to more than 2°C (3.6°F) above preindustrial levels. Many of these impacts involve a loss of biodiversity as species are unable to adapt the quickly changing climate and go extinct. In fact, recent research has concluded that there are signs we may be headed into a mass extinction event.[1]

A "mass extinction" event is characterized as a period during which at least 75 percent of the Earth's species die out in a geologically short interval of time. In the past 540 million years, only five such mass extinction events have occurred. If we continue on our present course, we could be headed toward a mass extinction event within a time frame of just a few centuries, although, fortunately, there is still time to reverse course.[2]

There is a very long history of discussions between international climate scientists in attempting to determine what level of global warming should be considered the danger limit which we should set as a target to avoid.[3] As a result of these discussions, 2°C (3.6°F) has become the internationally accepted danger limit target. If we continue on a BAU path, we will surpass the danger limit by the mid-21st century. In the best-case scenario we can still prevent the planet from warming beyond the danger limit, but only if we take major steps to reduce carbon pollution right away. This is strong evidence that we need to take action very soon to steer away from our current status quo emissions path.

So how do we achieve this transition away from our potentially catastrophic BAU path? The Australian government established a Climate Commission that released a three-chapter report entitled *The Critical Decade*, in part to help answer this question.[4] The Climate Commission suggested taking a budgetary approach in which humans are allowed a certain amount of carbon dioxide emissions over a given time frame. Their report suggested that humans should limit our emissions to 1 trillion tons of carbon dioxide between the years 2000 and 2050, in order to give ourselves a 75 percent chance to stay below the danger limit. Similarly, the International Energy Agency (an autonomous organization founded in response to the 1973 oil crisis, which conducts unbiased energy research and analysis) budgets approximately 1.2 trillion tons over that period to give us a 50 percent chance of staying below the danger limit.

A trillion tons may sound like a lot, but currently humans are releasing about 30 billion tons of carbon dioxide into the atmosphere every year. We've already burned through almost half the budget. At our current pace, it would take us only another 18 years to bust our 1 trillion-ton budget, and our emissions pace is currently going in the wrong direction. We're releasing more and more carbon pollution into the atmosphere ever year when we need to be cutting back. The year 2013 saw record high global carbon pollution emissions levels, at 37 billion tons. With the rapid development of the Chinese and Indian economies, it's also only going to get more difficult to get those emissions moving in the downward direction and get our carbon budget balanced by 2050.

The reason behind the Climate Commission's report title is that the longer we wait to make serious greenhouse gas emissions cuts, the harder it becomes to meet their proposed budget. The Climate Commission wrote:

> The peaking year for emissions is very important for the rate of reduction thereafter. The decade between now and 2020 is critical.

> Targets and timetables are, in principle, less important in the budget approach, but the urgency of bending emission trajectories downwards this decade implies that more ambitious targets for 2020 are critical in preventing delays in the transition to a low- or no-carbon economy.

For example, the Climate Commission concluded that if global greenhouse gas emissions had peaked in 2011, the annual emissions reduction rate would have to be no larger than 3.7 percent. However, if global emissions peak in 2020, the maximum emissions reduction rate jumps to 9 percent. In short, the longer we wait to reduce emissions, the more of our budget we eat through, and the steeper cuts will have to become in order to avoid breaking the budget by 2050. As *The Critical Decade* report concluded,

> The risks of future climate change—to our economy, society and environment—are serious, and grow rapidly with each degree of further temperature rise. Minimising these risks requires rapid, deep and ongoing reductions to global greenhouse gas emissions. We must begin now if we are to decarbonise our economy and move to clean energy sources by 2050. This decade is the critical decade.

In 2004, Princeton scientists Stephen Pacala and Robert Socolow published an influential paper in which they examined specifically how we can reduce greenhouse gas emissions using existing technology.[5] They developed the concept of "stabilization wedges," in which each wedge represents a certain amount of greenhouse gas emissions reductions. For example, if we replace enough coal power plants with wind turbines or solar panels, or build more fuel-efficient cars and drive less, these could each reduce emissions enough to represent one wedge. Each wedge represents a massive effort, like building 2 million large wind turbines. Pacala and Socolow came up with 15 ideas for stabilization wedges and argued that we need to achieve 7 of them by 2050 (although others have argued we need at least 14 wedges to avoid the danger limit).[6]

In September 2011, Socolow updated his work[7] and found that since his 2004 study, human greenhouse gas emissions had accelerated, increasing by nearly 30 percent over just a seven-year period. In order to achieve the same goal, Socolow concluded that we now must achieve nine of his wedges, rather than the original seven. Moreover, even if we achieve this new goal, the planet will ultimately warm roughly 0.4°C (0.7°F) more than it would have if we had started implementing the wedges in 2004, due to the emissions we've released in the meantime.[8] In short, Socolow's findings confirm the negative consequences of waiting another decade to act. Similarly, the International Energy Agency concluded in late 2011 that if there is no major international climate action by 2017, we will not be able to avoid the 2°C (3.6°F) danger limit.[9]

> If stringent new action is not forthcoming by 2017, the energy-related infrastructure then in place will generate all the CO_2 emissions allowed (. . .) up to 2035, leaving no room for additional power plants, factories and other infrastructure unless they are zero-carbon, which would be extremely costly.

In the United States, the critical decade has nearly became a lost decade. In our two-party political system, Republican politicians currently range between acknowledging that the planet is warming but refusing to do anything about it, and claiming that global warming is nothing but a massive hoax. Most Democratic Party politicians were afraid to even talk about global warming or climate change until just recently, but fortunately the tide finally seems to be turning, and more American politicians are starting to view tackling the threat of global warming as a winning political position.

CARBON PRICING IS THE KEY SOLUTION

The single most important step to solving the climate problem involves putting a price on carbon emissions. Right now across most of the United States, industries can release as much carbon pollution as they want for free. At least, the pollution is free for them. As with any kind of pollution, there is an unavoidable cost associated with greenhouse gas emissions. Those greenhouse gas pollutants are dumped into the atmosphere, which we treat like a giant landfill. They get mixed throughout the atmosphere and cause the global climate to change. This results in changing weather patterns, causing some types of extreme weather to become more intense and occur more frequently. Warmer ocean waters are fueling stronger hurricanes, and more powerful droughts, heat waves, and floods are battering farmers.

These climate changes have unavoidable costs, but when we don't pay for those costs up front when the pollutants are released, it's a problem known in economics as an "externality." When a cost associated with a product isn't reflected in its price, consumers are unable to make properly informed purchasing decisions, because they don't know the true costs of the product.

For example, let's say you pay $4 for a gallon of gasoline, but when accounting for the costs of climate change like more expensive food and increasing damages from hurricanes in the future, the actual cost of that gallon of gasoline is $5. That extra dollar of cost is spread around the world, paid by the farmers whose crops are damaged, by the people who have to pay higher food prices, by the people living near the coasts whose homes are flooded by higher hurricane storm surges, and so on. Somebody ultimately has to pay those costs, but you don't realize that when you fill up your gas tank, because the price of gasoline is artificially low. Hence, you're less likely to buy a fuel-efficient car, because gasoline prices seem cheap.

Economists hate externalities like this. If the market price of a product doesn't reflect its true cost, then the market can't work properly. Nobel Prize–winning economist Paul Krugman calls this basic Economics 101:[10]

> Externalities like pollution are one of the classic forms of market failure, and Econ 101 says that this failure should be remedied through pollution taxes or tradable emissions permits that get the price right.

British economist Nicholas Stern has similarly described the failure to put a price on carbon emissions as an incredible failure of the free market:[11]

> The problem of climate change involves a fundamental failure of markets: those who damage others by emitting greenhouse gases generally do not pay. . . . Climate change is a result of the greatest market failure the world has seen.

Indeed, there is nearly universal agreement among economists who study the subject as it relates to climate change that there should be a price on carbon emissions. Even Richard Tol—who is listed on the academic advisory council of the GWPF political advocacy group, which opposes taking action to address climate change, and who as previously discussed in this book attacked our paper showing a 97 percent consensus in the peer-reviewed climate literature that humans are causing global warming—has published research showing that there should be a price for carbon pollution.[12]

> A government that uses the same 3 percent discount rate for climate change as for other decisions should levy a carbon tax of $25 per metric ton of carbon (modal value) to $50/tC (mean value). A higher

> tax can be justified by an appeal to the high level of risk, especially of very negative outcomes, not captured in the standard estimates. . . . There is a strong case for near-term action on climate change, although prudence may dictate phasing in a higher cost of carbon over time.

Climate economists also agree that waiting to reduce our greenhouse gas pollution will cost us money in the long run, because it's cheaper to prevent climate change by reducing those emissions than it is to try and adapt to climate change once it happens. For example, prominent Yale economist William Nordhaus has written, in response to climate contrarians misrepresenting his work to argue we should wait 50 years to reduce our carbon pollution:[13]

> The cost of waiting fifty years to begin reducing CO_2 emissions is \$2.3 trillion in 2005 prices. If we bring that number to today's economy and prices, the loss from waiting is \$4.1 trillion. Wars have been started over smaller sums.
>
> My study is just one of many economic studies showing that economic efficiency would point to the need to reduce CO_2 and other greenhouse gas emissions right now, and not to wait for a half-century. Waiting is not only economically costly, but will also make the transition much more costly when it eventually takes place. Current economic studies also suggest that the most efficient policy is to raise the cost of CO_2 emissions substantially, either through cap-and-trade or carbon taxes, to provide appropriate incentives for businesses and households to move to low-carbon activities.

The failure to put a price on carbon emissions can also be viewed as an immense government subsidy. After all, it's a case of the government allowing companies to release greenhouse gas pollution into the atmosphere for free, despite the fact that we know those emissions have a cost via climate change impacts—costs that governments and taxpayers often have to absorb. And the size of that subsidy is breathtaking.

Economists try to estimate the costs of carbon emissions via climate change impacts in what they call Integrated Assessment Models. These models try to combine climate models and economic models to estimate the costs of climate change as best we can, which are known as the "social cost of carbon." PAGE09 is one example of an Integrated Assessment Model, run by Chris Hope at Cambridge University. The PAGE09 model estimates the average cost of a metric ton of carbon dioxide at about $100.[14] However, because there are significant uncertainties in both the size of future climate impacts and the costs of those impacts, there is a wide range of possible values for the social cost of carbon. It could be as low as just a few dollars per ton, and some economists argue it could be close to $1,000 per ton.[15]

Putting a price on carbon pollution is a tricky endeavor, because, for example, what price do you put on a human life? On potential cultural losses? On biodiversity and the value of species that may go extinct? Another moral and ethical issue is that poorer countries, which have contributed the least to climate change, tend to be the most vulnerable to its impacts.[16] This is due to the fact that poorer countries have fewer resources available to adapt to climate change and also because they often tend to be located near the equator (e.g., in Africa and Central America), in regions that are generally already hot and dry.

This is a key point that even climate contrarians tend to agree with. The Copenhagen Consensus Center, a think tank (despite its name, located in Massachusetts) run by Bjorn Lomborg who's best known for downplaying the threats posed by climate change, released a report in January 2014 on the costs of various global problems including climate change. The report's assessment of the costs of climate change was written by Richard Tol. Tol's climate assessment concluded that we've reached the point where further human greenhouse gas emissions and associated global warming will be worse for the global economy. Moreover, the report concluded that when "equity weighted" to better account for the impacts to the economies of poorer countries, we reached that point in 1980. Tol wrote:[17]

> Most countries benefitted from climate change until 1980, but after that the trend is negative for poor countries and positive for rich countries. The global average impact was positive in the 20th century. In the 21st century, impacts turn negative in most countries, rich and poor.

So even some climate contrarian groups recognize that climate change particularly harms poorer countries, who contribute the least to the problem. It's worth noting that Richard Tol's Integrated Assessment Model (called FUND) arrives at a significantly lower estimate of the costs of climate change than Chris Hope's PAGE09 model or William Nordhaus's DICE model. Nevertheless, they all agree that more global warming will be worse for the global economy and that the economies of poorer nations are being particularly hard hit by climate change.

The U.S. government currently uses a best estimate for the social cost of carbon of about $36 per ton of carbon dioxide. This a relatively low estimate but was raised in mid-2013 from the previous best estimate of about $22 per ton.[18] The problem is that although the U.S. government estimates that the cost of a ton of carbon dioxide pollution is $36, and despite the fact that economists agree that the cost of greenhouse gas pollution should be reflected in the costs of products that produce that pollution, right now American fossil fuel producers and consumers pay absolutely nothing for their carbon pollution (except in some regions like California that have put a price on carbon). The pollution is free; essentially a massive government subsidy.

Right now, global emissions from fossil fuels amount to over 31 billion metric tons of carbon dioxide per year. Using the U.S. government best estimate of the social cost of carbon, that amounts to a global subsidy of over $1 trillion per year. If we instead use the average estimate from Chris Hope's PAGE09 model, it's equivalent to an annual global fossil fuel subsidy of over $3 trillion. The United States alone emits about 5.2 billion metric tons of carbon dioxide pollution every year, amounting to an annual subsidy to the fossil fuel industry of $187 billion using the government best estimate, or $519 billion using the PAGE09 average estimate for the social cost of carbon. And that's just the indirect climate subsidies; governments also directly subsidize the fossil fuel industry, despite the fact that the industry has been established for well over a century and is the most profitable industry in the world.

According to estimates compiled by National Geographic using data from organizations like the International Energy Agency, direct government fossil fuel subsidies amount to around another $500 billion per year, globally.[19] That puts global fossil fuel subsidies conservatively at $1.5 trillion per year, going to fossil fuel companies like Exxon Mobil, which was the most profitable company in the world in 2011, making $41.6 billion in profits that year.[20] With such immense annual subsidies, it's no wonder that fossil fuel companies make such obscene profits, but why are we subsidizing the most profitable companies in the world?

CITIZENS CLIMATE LOBBY PUSHES FOR A REVENUE-NEUTRAL CARBON TAX IN THE UNITED STATES

There are a variety of different ways to put a price on carbon emissions. The most popular to date has been the concept of a carbon cap-and-trade system. Cap and trade involves putting a limit on the total carbon emissions that a given industry or trading system participants are allowed to emit, handing out or selling permits for those emissions, and then allowing participants to buy and sell them as needed. It's a free-market solution; originally a Republican concept, proposed as an alternative to pure government regulation of pollutants. As the Smithsonian Institute describes it,[21]

> The basic premise of cap-and-trade is that government doesn't tell polluters how to clean up their act. Instead, it simply imposes a cap on emissions. Each company starts the year with a certain number of tons allowed—a so-called right to pollute. The company decides how to use its allowance; it might restrict output, or switch to a cleaner fuel, or buy a scrubber to cut emissions. If it doesn't use up its allowance, it might then sell what it no longer needs. Then again, it might have to buy extra allowances on the open market. Each year, the cap ratchets down, and the shrinking pool of allowances gets costlier. As in a game of musical chairs, polluters must scramble to match allowances to emissions.

In the late 1980s, acid rain due to sulfur dioxide emissions was becoming a major concern. The George H.W. Bush administration pushed for a sulfur dioxide cap-and-trade system rather than simple government regulation and succeeded. According to the U.S. Environmental Protection Agency (EPA), by 2020, that system will have cost $65 billion, but the sulfur dioxide emissions cuts it will achieve will have created benefits (e.g., lower health care costs due to less air pollution) of $2 trillion and will have saved 230,000 people from premature death.[22] In 2010 alone, the acid rain program provided an estimated $120 billion in public health benefits—40 times the estimated cost. Despite this immense success from both environmental/public health and economic standpoints, Republican politicians universally claim that implementing a similar system for greenhouse gases will hamper the free market and cripple the U.S. economy.

The biggest downside to cap-and-trade systems is in their complexity and thus the potential for the system to be abused. A simpler option involves taxing carbon emissions. There are a number of details that need to be sorted out if a carbon tax is implemented. For example, should the tax be applied to the fossil fuel industries when they generate the products from which the greenhouse gas emissions originate, or should the tax be applied to consumers when the emissions are released? Most importantly, what should we do with the revenue generated by the tax? Some have suggested using it to reduce budget deficits, others think the money should be used to fund research and development of green technologies, while others think the money should be returned to the taxpayers.

The latter option is known as a "revenue-neutral carbon tax" or "fee and dividend," because rather than generate revenue for the government, the funds are all returned to the taxpayers. Again, there are a few different ways to make a carbon tax revenue-neutral. One option involves cutting other taxes by the same total amount as the carbon tax increase. The Canadian province of British Columbia (BC) took this approach, tying the carbon tax to reductions in personal and corporate income taxes, as well as tax credits to offset impacts on low-income individuals. In its first year (2008–2009), the system actually resulted in a net tax reduction of $230 million for BC residents, because the reduction in personal and corporate income taxes returned more money to the taxpayers than was generated by the carbon tax.

The main benefit of a revenue-neutral carbon tax lies in the fact that it minimizes the financial impact on taxpayers while nevertheless creating an incentive for individuals to reduce their greenhouse gas emissions. By taxing carbon pollution, BC citizens are encouraged to take steps to reduce those emissions, but by offsetting the carbon tax with reductions in other taxes, the average citizen will not experience any net tax increase. The BC system has been successful thus far, with per capita fuel usage falling more than 4 percent compared with the rest of Canada, and the BC economy has kept up with the rest of Canada's in the process.[23] The tax has also been very popular. Polls have shown that public support for the BC carbon tax has grown to 64 percent, and 59 percent of Canadians say they would support a similar carbon tax system in their provinces.[24] The popularity may be in part a result of the fact that by offsetting the carbon taxes, British Columbia has the lowest income taxes in Canada.

Another option involves simply returning the revenue generated by the carbon tax directly to the taxpayers, for example, with periodic rebate checks. Although the average cost to taxpayers is zero, the existence of the carbon tax nevertheless provides the incentive for people to reduce their consumption of fossil fuels and associated greenhouse gas emissions. In fact, those who reduce their emissions the most can come out ahead and make money from the revenue-neutral tax. Studies have shown that this will be the case for about two-thirds of taxpayers, while only the one-third with the highest carbon pollution will see their energy bills go up by a larger amount than the rebate checks they receive.

Most proposals involve increasing the size of the tax over time, so people know that the price of fossil fuels will only continue to rise, thus providing the incentive for taxpayers to reduce their fossil fuel consumption over the long term. Although the rebate checks will "make taxpayers whole," they will still have the financial incentive to reduce their carbon pollution. Who wants to waste his or her rebate on higher gasoline and energy bills?

In June 2014, an organization called Regional Economic Models, Inc. (REMI), published a report analyzing the economic impacts of a revenue-neutral carbon tax to various regions across the United States.[25] In a key conclusion of the report, because the refund checks will exceed the increased energy costs for about two-thirds of American taxpayers, REMI found that most people would actually have more disposable income as a result of this sort of carbon fee system. That increase in disposable income for people in most regions around the country would lead to increased consumer spending, which would translate into job growth and a stronger economy. The oil-heavy region consisting of Texas, Louisiana, Oklahoma, and Arkansas was the lone exception, with REMI finding that this region would take a modest economic hit, but the study found that every other region of the United States would benefit economically from a revenue-neutral carbon tax and that it would achieve substantial American carbon pollution cuts (with emissions declining by 33 percent after only 10 years and 52 percent after 20 years).

It's important to note that this report found that a revenue-neutral carbon tax would benefit the economy as compared to a baseline America without the carbon price. That baseline America did not account for the costs of damages from continued climate change if we don't reduce carbon pollution. The REMI study didn't consider climate change at all; it considered only the economic impact of the proposed change in the tax code if a revenue-neutral carbon tax were implemented. Thus, by failing to account for the economic damages we would avoid by slowing global warming, the report actually underestimated the economic benefit of the proposed carbon pricing system. In short, the evidence clearly points to a revenue-neutral carbon tax being good for the American economy.

Citizens Climate Lobby is a grassroots organization advocating for this type of revenue-neutral carbon tax. The most important aspect of this proposal is that it has significant support among American political conservatives. While a few Republicans like John McCain have supported carbon cap-and-trade systems in the past, despite originally being a Republican invention, there's never been enough Republican support to pass national carbon cap-and-trade legislation. In the United States, it's rare for legislation to successfully pass through Congress without significant bipartisan support.

However, because it's a free-market solution that doesn't create "big government," a revenue-neutral carbon tax appeals to conservatives and Republicans who understand the scientific and economic importance of putting a price on carbon emissions. The list of conservatives supporting a revenue-neutral carbon tax continues to grow:

- Fifty-one percent of Republican voters;
- Art Laffer, economic advisor to Ronald Reagan;
- Greg Mankiw, economic advisor to George W. Bush and Mitt Romney;
- George Shultz, Reagan's secretary of state and Nixon's Treasury secretary;
- Gary Becker, Nobel laureate in economics;
- Bob Inglis, former Republican congressman from South Carolina;
- William Ruckelshaus, EPA administrator under Nixon and Reagan;
- Lee Thomas, EPA administrator under Reagan;
- William Reilly, EPA administrator under George H. W. Bush; and
- Christine Todd Whitman, EPA administrator under George W. Bush.

The potential for bipartisan support and the minimal financial impact on American taxpayers are the main reasons that Citizens Climate Lobby specifically advocates for a revenue-neutral carbon tax. The group's strategy is to publish letters to the editor and opinion editorials in support of a carbon tax in newspapers around the country, and for individual chapters to meet with and encourage their local congress members to support revenue-neutral carbon tax legislation. The group has grown quickly, with chapters in most congressional districts across the United States, also expanding internationally. Citizens Climate Lobby members have met with hundreds of members of Congress, including Republicans like former House majority leader Eric Cantor.

The group also has a positive message, instructing member groups to highlight the positive actions of their congressional representatives when meeting with them or their staffers. Because of the group's effectiveness and rapid growth, it's been endorsed by a lot of big names in climate change, including James Hansen.

In mid-2013, I joined the Sacramento chapter of Citizens Climate Lobby. I think it has the right idea advocating specifically for a revenue-neutral carbon tax that conservatives can get behind. Our chapter has met with staffers in a few local congress members' offices, including both Democrats and Republicans. Citizens Climate Lobby executive Mark Reynolds and I published an opinion editorial in *The Sacramento Bee* in 2013, discussing that the science on human-caused global warming is settled, and we need to put a price on carbon emissions, preferably via a revenue-neutral carbon tax. I've also given several talks to local groups along with other Citizens Climate Lobby members, including to the Sacramento chapters of the United Nations Association and League of Women Voters in 2013, and at an interfaith discussion about the moral response to climate change in 2014, trying to build more local grassroots support for a revenue-neutral carbon tax.

If any group has a chance to break the Republican gridlock and obstruction on climate legislation, it's Citizens Climate Lobby. It's hard to see progress happening while any Republican in Congress who even accepts basic climate science and admits that the planet is warming will immediately be attacked by the right-wing media and will almost certainly face a primary challenge from a more extreme science-denying conservative candidate. However, Citizens Climate Lobby is working to change that political environment by reaching out to the general public with letters and editorials in newspapers around the country and by having positive, productive meetings with Republican members of Congress. Eventually moderates will once again be welcome within the Republican Party, and at that time a revenue-neutral carbon tax will become a real possibility in the United States. It's really a great solution that can grow the economy, create jobs, and give most people more disposable income, all while achieving serious cuts in carbon pollution.

STOPGAP EPA GREENHOUSE GAS REGULATIONS

In the meantime, the EPA has begun to implement regulations on greenhouse gas emissions from power plants during President Obama's second term. The process to establish these regulations actually began during the George W. Bush administration. The U.S. Clean Air Act has a provision that requires that for any air pollutant that contributes to air pollution and "may reasonably be anticipated to endanger public health or welfare,"[26] the EPA must set emissions standards. However, in 2003, the George W. Bush EPA announced that it would not regulate greenhouse gas emissions because Congress had not given the EPA the authority to regulate those emissions for climate change purposes, and because the EPA decided that setting those emissions regulations "is not appropriate at this time."[27]

As a result, 12 states (California, Connecticut, Illinois, Maine, Massachusetts, New Jersey, New Mexico, New York, Oregon, Rhode Island, Vermont, and Washington), three cities (New York, Baltimore, and Washington, D.C.), the territory of American Samoa, and a number of environmental groups took the EPA to court to challenge its decision not to regulate greenhouse gas emissions. The petitioners were represented by the Massachusetts Attorney General's Office. The case went all the way up to the U.S. Supreme Court, where it was known as *Massachusetts v. Environmental Protection Agency*. The U.S. Supreme Court made its decision on this case in 2007.

The main question at hand was whether greenhouse gases qualify as "air pollutants," in which case the EPA would be obligated under the Clean Air Act to determine whether they pose a threat to public welfare and thus whether their emissions must be regulated. This decision would be necessary only if greenhouse gases met the definition of air pollutants in the Clean Air Act:

> The term "air pollutant" means any air pollution agent or combination of such agents, including any physical, chemical, biological, radioactive (including source material, special nuclear material, and byproduct material) substance or matter which is emitted into or otherwise enters the ambient air.

Clearly this is a very broad definition, and greenhouse gases easily qualify. However, the petitioners led by the state of Massachusetts first had to prove that they had legal "standing," or the U.S. Supreme Court could have thrown the case out. To demonstrate standing, the petitioners essentially had to prove that they were being directly injured by the EPA refusal to regulate greenhouse gas emissions. The state of Massachusetts argued that as a coastal state, it was being adversely impacted by sea level rise causing the oceans to encroach onto its coastal properties.

This was a well-formed argument, because sea level rise is unquestionably directly connected to global warming. Sea level rise is caused almost entirely by two factors—melting land ice and thermal expansion (water expanding as it warms). Both are a direct consequence of global warming, and since we know that greenhouse gases cause the planet to warm, greenhouse gas emissions are directly related to sea level rise. Therefore, greenhouse gas emissions are injuring the state of Massachusetts by causing sea level rise that encroaches on the state's coastal property.

While this sounds like an open and shut case, in actuality it was very difficult to predict how the Supreme Court would rule on the issue of standing. In the 1970s, the U.S. Supreme Court was very liberal in its determinations of standing among environmental groups. An environmental organization essentially just had to argue that its members might visit an area that would be adversely impacted by a given action, and the Court would grant the group standing. However, over the past few decades, led by the conservative justice Antonin Scalia, the Court has gradually become more and more stringent in its determinations of standing for environmental groups. There had to be evidence that the group would be directly injured by the action in question. Though the state of Massachusetts had a strong case for injury and standing on greenhouse gas emissions, it was difficult to predict how the relatively conservative U.S. Supreme Court in 2007 would rule.

As is often the case, the Supreme Court split along politically partisan lines on the question of whether the state of Massachusetts had standing in this case. The four more liberal members of the Court (Stevens, Souter, Ginsburg, and Breyer) ruled in favor, while the four more conservative members (Scalia, Roberts, Thomas, and Alito) ruled against. And as is also often the case, it was left to the most moderate member of the court (Justice Anthony Kennedy) to break the tie and make the final decision. Kennedy correctly ruled that the state of Massachusetts had standing, and the majority decision required the EPA to make an "endangerment finding" to determine whether greenhouse gases endanger public welfare, in which case the EPA would be required to regulate their emissions.

For the following two years, the George W. Bush EPA essentially dragged its feet and delayed making this decision. When President Obama was elected and took office in 2009, the EPA quickly took up the task. The question was not a difficult one to answer. The body of scientific evidence and research is crystal clear in showing that human greenhouse gases are causing rapid global warming and climate change and that this climate change poses a threat to human welfare. For example, sea level rise caused by global warming threatens the coastlines of states like Massachusetts. More frequent and/or intense droughts and heat waves and floods endanger farm productivity, potentially causing food prices to rise. Warmer oceans are causing stronger hurricanes in the Atlantic, endangering the welfare of Americans in areas where hurricanes strike. We can't predict all of the impacts associated with climate change, but overall they certainly endanger public welfare.

The Obama EPA considered a number of major climate science assessment reports and correctly arrived at this same conclusion. In 2009 the EPA issued its greenhouse gas endangerment finding, concluding that[28]

> greenhouse gases in the atmosphere may reasonably be anticipated both to endanger public health and to endanger public welfare. . . . The major assessments by the U.S. Global Climate Research Program (USGCRP), the Intergovernmental Panel on Climate Change (IPCC), and the National Research Council (NRC) serve as the primary scientific basis supporting the Administrator's endangerment finding.

As a result, the EPA was required under the Clean Air Act to regulate greenhouse gas emissions. There were of course challenges to this decision. In February 2010, three states (Alabama, Texas, and Virginia) and several other parties sought judicial review of the EPA endangerment finding in the U.S. Court of Appeals, District of Columbia Circuit. However, two years later, the court dismissed the challenge. The three-judge panel unanimously upheld the EPA's finding that greenhouse gases endanger public health and welfare and are likely responsible for the global warming experienced over the past half-century.[29] Republican opposition to the EPA decision has continued, for example, with House Republicans attempting to pass legislation to defund the EPA in order to prevent them from enforcing regulations on greenhouse gases and other pollutants. They've also accused President Obama of overreach and abusing his executive powers, even though the carbon regulations are legally required under the Clean Air Act, as ruled by the conservative Supreme Court.

Nevertheless, the EPA has proceeded to issue greenhouse gas emissions regulations. They first issued regulations on emissions from vehicles, which are essentially met by the Corporate Average Fuel Economy standards, which require vehicles sold in the United States to meet a certain average fuel efficiency. These standards in turn limit the amount of gasoline they burn and hence the amount of greenhouse gases they emit.[30] In September 2013, the EPA issued a draft proposal for regulating carbon pollution from new power plants, taking into consideration more than 2.5 million comments from the public on its first such proposal, which was issued in 2012.[31]

The challenge was in setting standards on both coal and natural gas power plants, because the former have much higher carbon dioxide emissions than the latter. The EPA had originally proposed to create one set of emissions standards, which natural gas power plants would be able to meet, but which would be nearly impossible for coal power plants to meet. Fearing that the coal industry might successfully be able to challenge these regulations in court as unfair to coal power plants, the new 2013 EPA proposal set different greenhouse gas emissions standards for natural gas and coal power plants. Nevertheless, it's expected to be difficult and expensive for new coal power plants to meet these standards, requiring them to capture and store some of their carbon emissions, which is a costly process. Construction of new coal power plants had already stalled, as they've been replaced by cheaper natural gas and renewable energy power plants instead. The addition of these new EPA greenhouse gas regulations makes it unlikely that many new coal power plants will ever be built in the United States. In June 2014, the EPA issued its draft proposal for regulating carbon pollution from existing power plants as well.

The great irony is that by forcing President Obama's hand, Republicans have given him no choice but to implement "big government regulations" when they could have instead supported a free-market solution. After being elected to a second term, in his February 2013 State of the Union Address, President Obama told Congress that if it failed to pass "bipartisan, market-based" climate change legislation to "protect future generations, I will."[32] A few months later, in August 2013 his secretary of state John Kerry reiterated this climate commitment to his Brazilian counterpart:[33]

> So the challenge is ahead of us, for all of us, and I know that the United States has a great commitment under President Obama to take our own initiatives, not even to wait for congressional action, but to move administratively in order to do our part. I know we can continue to work with Brazil on this issue of climate, and we look forward to doing so.

The Obama administration had made it abundantly clear that it would prefer for Congress to address the climate change problem through a market-based solution like a carbon tax or cap-and-trade system, but that if it failed to do so, his EPA would implement the legally required government greenhouse gas regulations. Republicans in Congress knew that these were their two options, that government regulations were all but inevitable if they failed to implement free-market climate legislation, and yet they still failed to do so. They have continued to obstruct policies supported by conservatives that would be better for the economy than government regulation of carbon pollution. Nevertheless, Republicans could replace these regulations with a small government revenue-neutral carbon tax at any time. That policy already has support among Democrats; all that's needed is a little support from Republicans for this free-market alternative.

THE POLITICIZATION OF SCIENCE

Unfortunately, the political atmosphere in the United States has become so toxic and partisan, and the Republican Party has so politicized science, that it's simply not possible for most Republican policymakers to support any sort of climate policy, even if it embodies a free-market solution replacement to big government regulation. Doing so would guarantee heavy criticism from the conservative media and a primary election challenge by a candidate from the extreme right wing of the party. Bob Inglis is a perfect example of this toxic, partisan, antiscience political climate in today's Republican Party.

Bob Inglis was the Republican congressional representative from South Carolina's 4th District from 1993 to 1999 and from 2005 to 2011. Inglis was a conservative congressman from a conservative district in a conservative state; he had a 93.5 percent lifetime rating from the American Conservative Union and had been endorsed by the National Rifle Association and National Right to Life. However, Inglis was also a realist when it came to the subject of climate change and supported taking action to address the problem. In 2010, he was challenged in the Republican primary by a Tea Party–backed candidate, who portrayed Inglis as not sufficiently conservative, despite his high conservatism rating from the American Conservative Union. Some of Inglis's other positions were challenged, like his opposition to the proposed 2007 Iraq War troop surge, but his acceptance of the expert consensus position on human-caused global warming was one of the key issues on which Inglis was portrayed as insufficiently "conservative." Inglis lost the primary election to the Tea Party candidate by a landslide, 71 percent to 29 percent.

That election result may have spooked other Republican members of Congress around the country, because few if any have since expressed any support for tackling the problems posed by climate change, and in fact few will even admit in public that human-caused global warming is a scientific reality. Fear of a primary challenge by a more conservative candidate is pervasive among Republicans in Congress. Sadly, their political party has come to view climate change as a political issue rather than a scientific one.

After he lost his seat in Congress, Bob Inglis started a nationwide public engagement campaign promoting conservative and free-enterprise solutions to energy and climate challenges, called the Energy and Enterprise Initiative.[34] Inglis has become an outspoken, staunch supporter of a revenue-neutral carbon tax and hence has also been very supportive of Citizens Climate Lobby. Inglis joins the long list of conservatives who support a free-market solution to global warming, but because of the partisan, antiscience position of today's Republican Party, they have yet to convince any conservative members of Congress to sponsor or publicly support a revenue-neutral carbon tax or other climate legislation. Instead, the Republican Party has obstructed all congressional efforts to address the threats posed by climate change.

As a result, President Obama was forced to instead use his executive powers to address the problem, as he had promised, and as was legally required. In September 2013, Obama's EPA released its draft rules to regulate emissions from coal-fired power plants and set up a climate change adaptation task force. In early December he ordered federal agencies to lead by example and get 20 percent of their power from renewable sources by 2020. He established a Climate Action Plan, with EPA greenhouse gas regulations as the featured component.[35] The plan also involves increasing fuel economy standards to reduce greenhouse gas emissions via transportation, improving energy efficiency in buildings, reducing other greenhouse gas emissions besides carbon dioxide, preserving forests, leading on the issue of climate change at the federal level and in international negotiations, and preparing to adapt to the climate change that we can't avoid.

Aside from the concrete steps to reduce greenhouse gas emissions, President Obama's actions have been important for the message they've sent. In his first term in office, President Obama took some relatively small and low-profile steps to address the problem, but rarely spoke about climate change. When faced with the choice to spend his political capital on health care or climate change, he chose health care. It was a disappointment to those who voted for him in the hopes that he would take a leading role in tackling global warming. Arguably what we most lack are political leaders who are willing to speak out about the importance and urgency of addressing climate change, which is one of the main reasons why the public doesn't view addressing global warming as a high priority.

However, President Obama has made a significant shift on climate change in his second term in office. He began speaking about the issue in prominent speeches, like in his second inaugural address and State of the Union. He appointed John Kerry, who has long been a strong advocate of addressing climate change and has coauthored bipartisan climate legislation in the U.S. Senate, as his secretary of state. He promised to use his executive powers to address climate change if Congress failed to do so, and he followed through on that promise. The United States has also begun taking a leadership role in international climate negotiations, whereas during the George W. Bush administration, the United States was a key roadblock to making any significant international progress.

Unfortunately, the other half of the American political landscape continues to try and block all progress in solving the threat posed by climate change. In his January 2014 State of the Union address, President Obama said:

> Climate change is a fact. And when our children's children look us in the eye and ask if we did all we could to leave them a safer, more stable world, with new sources of energy, I want us to be able to say yes we did.[36]

It's encouraging to hear the president speak out in support of climate science, but simultaneously discouraging that he's forced to state something as obvious as "climate change is a fact." Even more discouraging, while these comments made by President Obama on climate change received strong applause from Democrats in the audience, they were predominantly met with silence from the congressional Republicans in the room. It was another sign of the conservative politicization of science.

PROGRESS AT THE MORE LOCAL LEVEL

Fortunately, there's been some good news on more local levels. In recent years, some states have stepped up to implement measures to reduce local greenhouse gas emissions. California has begun implementing the most aggressive emissions reduction system in the United

States, with a target of reducing the state's greenhouse gas emissions to 1990 levels by 2020 and 80 percent below 1990 levels by 2050.[37]

To achieve this aggressive goal, California has launched a carbon cap-and-trade system, which essentially uses the free market to put a price on greenhouse gas emissions, and thus gives industries financial incentive to reduce their emissions.

Australia also approved a carbon tax that was implemented in mid-2012 and was planned to transition to a cap-and-trade system in three to five years.[38] Unfortunately, the Labor Party, which had implemented the carbon pricing legislation, lost popularity for various reasons. The carbon price was controversial, but most Australians supported it going into the September 2013 election, won by Tony Abbott's Liberal Party (which despite the name is politically conservative).[39] Polling immediately after the election showed that the carbon tax was the main issue deciding their vote for only about 3 percent of Australian voters.[40] Nevertheless, Australia took a big step backward by repealing the carbon tax in July 2014.

In 2014, China completed the rollout of seven regional carbon cap-and-trade systems.[41] The Chinese are evaluating whether to pursue a cap-and-trade system, carbon tax, or regulatory limits to reduce its carbon pollution. A decision is expected by 2018, when the chosen policy is expected to be implemented. China is often used as a scapegoat by those looking for an excuse to oppose American action to tackle global warming. Because of China's large population and rapidly growing economy, it produces more total carbon pollution than the United States, although significantly less per person. Thus, until China tackles its carbon emissions, opponents argue, there's no sense in America tackling its carbon emissions. However, as these regional carbon cap-and-trade systems and plans for a national policy within a few years show, China is actually ahead of the United States in addressing the problem domestically.

VERY SLOW PROGRESS AT THE INTERNATIONAL LEVEL

At the end of 2011, another round of international climate negotiations were held in Durban, South Africa. The Durban talks followed major international climate conferences in Kyoto in 1997, which resulted in the Kyoto Protocol (where most developed nations, except the United States, agreed to reduce their greenhouse gas emissions), and in Copenhagen in 2009, where little progress was made. Expectations for the Durban negotiations were very low, but the good news is that some progress was made. Both developed (including Europe and the United States) and developing (most notably China and India) nations agreed on a framework to establish legally binding greenhouse gas emissions reductions targets, in an extension of the Kyoto Protocol. The bad news is that these targets don't need to be implemented until 2020, and the longer we wait to start reducing emissions, the more painfully large the cuts will have to be. However, while progress is slow, at least progress is being made, which is critically important.

Another round of talks were held in Warsaw, Poland, in 2013. The negotiators agreed to kick the can down the road once more; countries have until the first quarter of 2015 to publish their plans to curb their greenhouse gas emissions. The goal of the Warsaw conference was to lay the groundwork for the next international negotiations, to be held in Paris in late 2015. The 2015 Paris conference is now seen as a critical one, where international negotiators must agree on concrete plans to reduce carbon pollution and slow global warming.

LOW-CARBON TECHNOLOGY SOLUTIONS

The good news is that we have all the technology we need to solve the problem. Wind turbines are a particularly cheap, cost-effective, low-emissions energy source. In the United States, according to the Energy Information Administration, the states generating the highest percentage of their energy supply from wind as of 2011 were South Dakota (22.3 percent), Iowa (18.8 percent), North Dakota (14.7 percent), Minnesota (12.7 percent), Wyoming (10.1 percent), Colorado (9.2 percent), Kansas, Idaho, and Oregon (each with 8.2 percent).

These are generally not liberal, hippie, tree-hugger states. They're installing all of these wind turbines for purely economic reasons; it's a cheap source of energy. It helps that the midwestern United States is often referred to as "the Saudi Arabia of wind." There's an area splitting the country almost right down the middle, from North Dakota to Texas, where wind speeds are consistently high, helping wind turbines generate a lot of energy and hence making them particularly cost-effective. Overall, U.S. wind generation has grown by over 140,000 megawatt-hours since 1997. In 2013, wind generation surpassed 4 percent of total electricity generation in the United States, up from less than 1 percent just six years earlier. It's still a small percentage, but wind energy production is growing fast.

Solar panels are rapidly falling in price and growing in popularity as well. The installed price of solar panels in the United States fell by about half between 1998 and 2012. The installed energy-generating capacity of solar photovoltaics in the United States doubled from 2007 to 2009, and then nearly doubled again each year from 2009 to 2010, 2010 to 2011 to 2012, and 2012 to 2013. Between 2008 and 2013, the amount of new solar photovoltaic installations in the United States increased by a factor of 10. California is leading the way, accounting for about 35 percent of American solar photovoltaic energy capacity, followed by Arizona (15 percent) and New Jersey (13 percent).

California has also begun deploying another type of solar energy production called concentrated solar thermal power. These systems use mirrors or lenses to concentrate sunlight onto a small area, like a central tower. The concentrated light is then generally used to heat a fluid like water or molten salt. The heat is then used to drive a steam turbine to generate electricity. These concentrated solar thermal systems can also provide what's known as "baseload" power—energy that can be produced at any time—because the liquid can also be used to store the heat generated by the concentrated sunlight. One criticism of wind and solar power is that they only generate electricity when the wind is blowing and the sun is shining (and hence don't provide baseload power, unlike fossil fuels for example, which we can burn at any time), but concentrated solar thermal plants can store energy for many hours and thus can potentially be used as a baseload power source.

Spain has been the global leader in concentrated solar thermal energy development and deployment, but California has begun constructing solar thermal power plants as well. In the Mojave Desert, the Ivanpah Solar Electric Generating System was completed in 2014. This is the world's largest concentrating solar facility, generating power by using over 170,000 mirrors to concentrate sunlight on a central power tower. The Ivanpah solar plant is expected to generate over 1 million megawatt-hours of energy per year, which is enough to power over 140,000 homes in California.

We have a very long way to go before solar energy makes a dent in overall national or global energy production—it still produces less than 1 percent of the electricity consumed in the United States—but the rate at which its costs are falling and installations are growing is remarkable and encouraging. The number of people employed by the solar industry is growing rapidly as well, with one of the fastest rates of growth of any American industry. The solar industry employed 119,000 people in 2012, up 13 percent from 2011.

The necessary resources are also readily available. The southwestern United States has plenty of sunny, open land, and the Midwest has an abundance of areas with high wind speeds. In fact, to meet the entire American electricity demand with concentrated solar thermal power would require an area of just 92 square miles in size. American cities and residences cover about 140 million acres of land, and we could supply all of the country's current electricity requirements by installing solar panels on 7 percent of this area (e.g., on roofs, on parking lots, and along highway walls). So we have the technologies to produce low-emissions energy, and we have more than enough renewable resources (wind and sunlight and hydroelectric) to produce all the electricity we need. The costs are falling as the technologies improve and become more widely used. All we need is the will to continue deploying these technologies to replace electricity generated by fossil fuels.

Research has also shown that when all costs are taken into account, renewable energy is often cheaper than energy from burning fossil fuels. A 2012 paper written by Laurie Johnson of Natural Resources Defense Council and Chris Hope of Cambridge University compared the costs of various types of energy when accounting for the climate costs associated with their greenhouse gas emissions.[42] It's important to remember that carbon pollution has very real costs that can't be avoided. For example, the global warming caused by greenhouse gas emissions from burning fossil fuels leads to more and stronger heat waves, which in turn are bad for crops. Lower agricultural yields lead to increased food prices. Global warming also causes sea level rise, and warmer oceans tend to fuel stronger hurricanes, both of which increase the damage caused by these storms. Accounting for these costs when carbon pollution is emitted is the purpose of the aforementioned social cost of carbon.

Johnson and Hope found that as of 2012, wind energy was already cheaper than coal without even considering these climate costs. They also found that solar photovoltaic energy costs would be break-even with coal energy costs if the social cost of carbon is $50 per ton. Moreover, the cost of wind energy would be as low as energy from natural gas if the social cost of carbon is about $74 per ton. Recall that the U.S. government best estimate for the social cost of carbon is $36 per ton, but according to Chris Hope's research, the best estimate is $100 per ton. In short, wind is certainly one of the cheapest forms of energy available, and solar photovoltaic energy is already close to the true cost of coal, with prices falling rapidly.

The positive and rapid growth trends in solar energy are aided by innovative companies like Solar City, chaired by the incredibly successful entrepreneur Elon Musk (who some have claimed was the model for the character Tony Stark in the *Iron Man* films). Solar City and some other similar solar companies have bypassed the problem posed by the high up-front capital costs associated with buying solar panels by allowing their customers to instead lease the panels. Some of these companies give their customers the option to choose between paying a monthly fee, paying for a long-term lease (e.g., for a 10-year period) up front at a lower cost, or a combination of these options.

In 2010, I began leasing solar panels from a California company called Sungevity. In order to get the best deal, I paid for a 10-year lease up front. The average cost over the 10-year period is about the same as I would have paid my local electric utility, but instead I'm producing my own carbon-free renewable energy. Each year the solar panels on my roof are producing more than 3 megawatt-hours of electricity, which was initially more than my household energy consumption. What I don't consume goes back into the electric grid, and my electrical utility has actually had to send me a check to pay for that extra electricity my solar panels have produced for them.

That changed in late 2011, when I was able to purchase another low-carbon technology—an electric car (a Nissan Leaf). Research has shown that electric cars produce lower greenhouse gas emissions even when the electricity fueling them comes mostly from coal-fired power plants, in large part because electric motors are far more efficient than gasoline internal combustion engines. A study by the Union of Concerned Scientists found that in regions covering a total of 45 percent of the American population, electricity generation is clean enough that electric cars would produce lower greenhouse gas emissions than even the most fuel-efficient hybrid cars.[43] In regions that get most of their electricity from burning coal, hybrid cars produce lower greenhouse gas emissions than electric cars, which would be on about even footing with standard gasoline-powered cars in those coal-heavy regions. However, when charged up from a renewable energy source, an electric car can travel on essentially zero greenhouse gas emissions. Combining electric car ownership with solar panels ensures that it's the lowest emission form of individual road transportation, aside from human power like bicycling.

There are of course still roadblocks to widespread adoption of electric cars. Batteries are becoming cheaper as the technology improves, but a relatively affordable new electric car like the Nissan Leaf gets only about 70 miles per charge in real-world driving and then requires many hours to fully recharge the battery. Another of Elon Musk's companies, Tesla Motors, has made significant strides in solving this problem. Its Tesla Roadster and Model S electric vehicles can get upward of 250 miles per charge, and the company is installing superfast, free recharging stations all over the country, which let customers charge their batteries most of the way up in about an hour, at no cost.

While free fuel is a major perk of owning a Tesla electric vehicle, the sticker price is still a barrier to most people, at upward of $70,000. The good news is that Tesla is aiming to start selling a more affordable electric vehicle beginning in 2016 or 2017, with a target price of around $40,000 and a target range of about 200 miles per charge.[44] With the possibility of federal and state tax breaks to encourage the purchase of low-emissions vehicles, this could make a long-range electric car affordable for a significant proportion of the American public. It's only a matter of time before electric vehicles become the norm, and gasoline-powered vehicles are perceived as Stone Age technology.

In the meantime, plug-in electric cars offer a practical compromise. These are cars with both internal combustion engines and electric motors. For the first 40 miles or so, plug-in hybrids run on the electric motor and batteries, just like an electric car. After the batteries run out of most of their charge, these cars switch to a traditional hybrid mode, using a combination of the internal combustion engine and gasoline fuel, and the electric motor and batteries. This is a good compromise because it offers the efficiency and low emissions of an electric car for short trips, but also includes the long range ability of hybrids or standard gasoline-powered cars. The Chevy Volt is the most popular plug-in hybrid on the market, with a 40-mile electric range. The Volt and the Nissan Leaf are in competition to be the most popular car with an all-electric option. In 2013, Chevy sold 23,094 Volts, while Nissan sold 22,610 Leafs. In 2014, the Leaf began to surge past the Volt in popularity and sales.

THE BOTTOM LINE

The bottom line is that the global warming predictions that have been the most accurate foresee very dangerous and potentially catastrophic climate change in the near future if we continue with our current path of heavy reliance on fossil fuels. We're not talking about several generations from now; we're currently on track to pass the global warming danger limit within our or our children's lifetimes. Optimistic predictions to the contrary have already proven to be inaccurate just a few years after they were made, and yet a significant number of people continue to listen to the contrarians despite their history of wrong assertions and predictions. They should no longer be considered credible sources of information about climate change.

To avoid blowing past the danger limit, we need to start taking serious steps to reduce our greenhouse gas emissions. This will involve transitioning away from fossil fuel dependence toward renewable energy sources, alternative fuel technologies like electric cars, improving the energy efficiency of our homes and other buildings, and generally consuming less energy. It will also require implementing some sort of system to put a price on carbon emissions. Right now in the United States, we're allowed to pump out as much carbon pollution as we want for free, but even though the costs of climate change are not reflected in the market price of energy, we cannot avoid paying those costs.

Some local and national governments have taken steps to put a price on carbon emissions, but overall our response has been inadequate for the task at hand. Much of the fault lies with the United States in particular, which is responsible for nearly 30 percent of the increase in atmospheric carbon dioxide over the past 150 years, nearly triple China's share (the second-largest contributor to atmospheric carbon dioxide), despite the fact that China has four times the population of the United States.[45]

As the world's largest historical emitter, and arguably the world's biggest superpower, the United States should be taking the lead by reducing its carbon pollution. Instead, it has been one of the slowest nations to act and has become a bastion of climate contrarianism and obstructionism. A large chunk of Americans are woefully misinformed about climate science (particularly the most politically conservative, especially those who self-identify as Tea Party members[46]). Unfortunately, climate science is viewed as a political rather than scientific subject by many Americans, who allow their political ideologies to cloud their perceptions when it comes to climate change. We can only hope this changes in the near future before it becomes too late to avoid catastrophic consequences.

Fortunately, President Obama's Climate Action Plan and EPA greenhouse gas emissions regulations have been big steps in the right direction. California is once again leading the country on environmental issues with its own carbon cap-and-trade system. Citizens Climate Lobby is building up bipartisan grassroots support around the country and around the world for an economically beneficial revenue-neutral carbon tax, and the organization is growing at an incredible rate, doubling in size and productivity every year. Countries like China and

Mexico are taking big steps to reduce their greenhouse gas emissions, and European countries have long been leaders in tackling climate change.

We also already have all the technology we need to stop global warming. If we could just muster the will, all of the world's energy could be generated from clean low-emissions renewable sources like wind, solar, tidal, and geothermal power. Nuclear power may need to be a significant part of the mix as well, although at present, new nuclear power plant construction projects have a difficult time competing economically because they usually go well over schedule and budget, and American taxpayers are on the hook for the very expensive bill if the projects default on their loans. However, if we want to avoid the most dangerous climate change consequences, all viable solutions need to be on the table, including nuclear power. These low-emissions energy sources can be used to fuel electric vehicles in order to produce clean transportation. Technologies and buildings are becoming more energy efficient, which has the added benefit of saving people money.

All we need is the will to deploy these climate solutions. It would be an easier pitch to sell if the prices of products reflected their true costs, including the costs of the climate damages they cause. That will require putting a price on carbon pollution. Some countries and regions have accomplished that, but while EPA greenhouse gas regulations are an important step, a carbon price in the United States would be an even bigger and better one. If the costs of carbon pollution were reflected in the prices of products, it would give people the financial incentive to reduce their greenhouse gas emissions. If 100 percent of the revenue generated from that carbon tax were returned to the taxpayers, it wouldn't hurt people financially, and in fact about two-thirds of Americans would receive a rebate check that would more than offset their higher energy costs—they would come out ahead. Only the biggest polluters would pay more than they would receive back in rebates, and the system would be good for the economy as a whole.

Denial and ideological biases are the main roadblocks to the deployment of these climate solutions. Fortunately, the rejection of climate science is mostly only a partisan issue in the United States. Other countries whose governments and conservative political parties fail to take the problem seriously, as in Canada, at least pay lip service to climate change. Australia is becoming another exception to the rule, with the Liberal Party under Prime Minister Tony Abbott outright denying and rejecting basic climate science more and more as time goes on. Nevertheless, partisan rejection of climate science is relatively rare outside of the United States.

Our biggest roadblock to solving the problem lies in the fact that most people don't recognize the urgency of the problem. The longer we wait to reduce greenhouse gas emissions, the steeper our cuts will need to be to stay within our allowable carbon budget and prevent extremely dangerous global warming. And the longer we wait, the more expensive it will be. If we wait too long, eventually it will become infeasible to avoid the extremely dangerous climate change, and we'll be stuck devoting most of our resources to limiting the damage as much as possible and paying the immense costs of damages we can't avoid. Time is running out.

Most people don't recognize this urgency because they don't realize there's an expert consensus on human-caused global warming. When asked how many climate scientists agree that humans are causing global warming, the average American answer is 55 percent, while the reality is 97 percent. This consensus gap is a critical problem, because when people realize there's an expert consensus on the issue, research has shown they're more likely to accept the science and support taking action to address the problem.

I believe the main cause of the consensus gap is due to false balance in media coverage, with the less than 3 percent of climate contrarians receiving disproportionate media coverage. Some journalists and news outlets are guilty of this false balance because they're pandering to a politically conservative audience, and climate science is treated as a politically partisan issue. A paper published in the *Journal of Communication* in September 2014 found the following:[47]

> Thus, by demonstrating that media use not only reinforces certainty or uncertainty in global warming but also, in turn, reinforces support or opposition for policies to mitigate global warming, our results point to the important role of the media in advancing—or hindering—policymaking related to global warming. Specifically, our results suggest that governmental inaction on climate change can partially be attributed to the echo chamber created by conservative media on the issue.

Other journalists and media outlets are guilty of this problem because they don't understand what real balance entails. Some media outlets think that every story needs to be "balanced" with views from "both sides" and that this sort of "balance" is a sign of good unbiased journalism. In reality, if more than 97 percent of experts are on one side and less than 3 percent are on another, giving them equal weight misleads the audience. It's a journalistic failure.

Another problem is that controversy sells in the media. People are unlikely to read a story about yet another study or scientist confirming the 97 percent consensus that humans are causing global warming. However, a rogue scientist or study that claims to overturn everything we know about how the world works makes for a juicy story. The problem is that it's rare for a single scientist or study to overturn our understanding of an entire scientific field. As the great scientist and communicator Carl Sagan said,

> They laughed at Columbus, they laughed at Fulton, they laughed at the Wright brothers. But they also laughed at Bozo the Clown.[48]

Most of the climate contrarians who are the focus of so much media attention bear much more resemblance to Bozo the Clown than the Wright Brothers. The easy way to make that determination is to look at the accuracy of their predictions, as I've done in this book. Richard Lindzen is my favorite example—a contrarian climate scientist who's automatically considered credible by journalists because he was employed by MIT (of course, these journalists never mention that nearly every other climate expert at MIT strongly disagrees with Lindzen). Entire stories have been centered around Lindzen, and he has been invited to testify before the U.S. and UK governments. Yet Lindzen has been proven wrong about nearly every major climate statement he has made. He claimed that the planet hadn't warmed significantly, that instead the temperature record was wrong, that it wouldn't warm noticeably in the future, that changes in clouds and water vapor would act to dampen global warming, and so forth. Wrong, wrong, wrong, wrong, wrong on every count. As shown in this book, climate contrarian predictions of global cooling or minimal global warming have also proven consistently wrong.

Yet even after they join politically driven organizations, Lindzen and his fellow contrarian outliers are still treated as credible experts by many in the media despite their long histories of being constantly proven wrong. And as long as they continue to tell conservative media outlets what they want to hear, and continue to provide "the other side" for journalists who can't tell the difference between real balance and false balance, they will continue to be treated as credible experts.

And that's a major problem. Our lack of action to address what may very well be the biggest threat humanity has ever faced can be traced directly back to this journalistic malfeasance.

Politicians don't need to support policies to address climate change because their voters don't see it as an urgent issue. The voters don't see it as an urgent issue because they perceive there's still substantial scientific debate and disagreement on the subject (the consensus gap). And the consensus gap exists because the fringe minority of contrarians are given disproportionate coverage in the media under the guise of "balance."

However, it's not considered good journalism to "balance" a discussion about evolution with the opinions of a Creationist, or to talk to a flat Earther after interviewing an astronaut, or to follow a discussion on the health effects of smoking by bringing on an "expert" who claims cigarettes don't cause cancer. The latter example is particularly appropriate, because many of the same voices who denied or downplayed the link between smoking and health cancer now deny or downplay the links between human greenhouse gas emissions and global warming. Richard Lindzen is again a good example, having disputed the confidence with which we know secondhand smoke causes lung cancer.[49] Many who now work at political think tanks denying that humans cause global warming previously worked at political think tanks denying that smoking causes lung cancer, using the same tactics in both campaigns of disinformation.

In fact, the level of confidence among climate science experts that humans are the main cause of global warming is as high as the level of confidence among medical science experts that smoking causes cancer. The question is, when will it be considered bad journalism to "balance" the consensus view on climate science with fringe contrarian view the way it's considered bad journalism to balance the consensus view on smoking and cancer with a fringe contrarian view? When will the contrarians' history of being wrong about global warming finally undermine their credibility in the media? Former acting assistant secretary of the U.S. Department of Energy and climate blogger Joe Romm accurately described the failure of many of today's journalists to consider the credibility of their climate sources:[50]

> A defining characteristic of modern journalism is a lack of judgment, an unwillingness—or inability—to disqualify anyone as a credible source on a subject no matter how thoroughly discredited they have been by reality.

Unfortunately, as long as the media views climate science as a political issue, contrarians will likely continue to be considered credible alternative voices, and the public will continue to be misinformed. We need political leadership to trump this journalistic malpractice so that the public comes to understand that climate contrarians represent fringe outlier views that are not supported by the full body of scientific evidence. President Obama and a number of his colleagues in the Senate and House of Representatives have begun taking a leadership role on climate change. Senator Sheldon Whitehouse of Rhode Island has been giving weekly climate speeches on the floor of the Senate, and in March 2014, over 30 Senate Democrats stayed up all night in a marathon-speaking session devoted to climate change.

In the Republican Party we've seen some great leadership on climate from former representative Bob Inglis of South Carolina, economic advisor to George W. Bush and Mitt Romney Greg Mankiw, President Ronald Reagan's secretary of state George Shultz, former Republican heads of the EPA, and many others. However, climate leadership among Republicans currently holding office in Congress has been virtually nonexistent. Groups like Citizens Climate Lobby are working hard to change this, but so far Republicans continue to view climate change as a partisan issue and fear that if they do the right thing, they'll face a primary challenge from an extreme science-denying opponent and lose their jobs.

We need more courageous leaders in the Republican Party like Bob Inglis who are willing to do the right thing even at the risk of losing their seats in Congress. There's really no reason why climate change should be viewed as a partisan issue. Science has nothing to do with politics, and there are free-market, small government solutions like a revenue-neutral carbon tax that many conservatives support. All we need is a bold and courageous Republican to take the lead on this issue and pull his or her party out of the science-denying dark ages. That person would undoubtedly go down in the history books as hero.

It will inevitably happen eventually. As extreme weather events become more frequent and more damaging, more people will come to accept the dangerous reality of human-caused global warming. Science always wins out in the end because it's based on physical realities, and a political platform that denies science cannot last. As astrophysicist and brilliant science communicator Neil deGrasse Tyson says,

> The good thing about science is that it's true whether or not you believe in it.[51]

The only question is how long the fossil fuel industry-funded science denial can continue to misinform the public and buy political favors. Younger generations are becoming more knowledgeable about climate change and aware that they'll face the brunt of the impacts caused by the choices of older generations. If Republicans continue to deny climate science and oppose solutions, they'll lose many of those voters forever.

It may only take one visionary Republican politician to understand that the party must alter its position on climate change. If one such politician can succeed, it will break the monolithic climate denial among party leaders and prove that scientific realism isn't a political death sentence in the Republican Party. As President Obama showed when he changed his position to come out in favor of gay marriage, when leaders lead, their supporters often follow. We also need people of all political stripes to help build the grassroots support that will allow a courageous Republican leader to take a stand in favor of science and protecting future generations. On that front, I can't say enough about the great work Citizens Climate Lobby is doing and how proud I am to be part of the organization.

We're very close to solving the climate problem. We have the technology needed to do it, we know the main policy we need to implement (a revenue-neutral carbon tax), we can avoid the most dangerous consequences if we act soon, and we're building bipartisan support. If the media will just hold contrarians accountable for their history of being wrong and stop misinforming people with false balance in climate reporting, we may still be able to prevent the worst climate consequences. All we need is an adequately informed public and the will to act. We're very close to taking the path toward ensuring a stable, livable climate for future generations, but the window for us to take that path is closing.

Notes

PREFACE

1. E. Steig. Berkeley Earthquake Called Off. http://www.realclimate.org/index.php/archives/2011/10/berkeley-earthquake-called-off/.

CHAPTER 1

1. Spencer Weart. The Discovery of Global Warming. http://www.aip.org/history/climate/index.htm.
2. Skeptical Science. Rebuttal to "CO_2 Is Just a Trace Gas." http://www.skepticalscience.com/CO2-trace-gas.htm.
3. G.D. Thompson. What Do You Know about CO_2 and Climate Change? http://www.theonelightgroup.com/latest/what-do-you-know-about-co2-and climate-change.
4. No Carbon Tax Rally. The Truth about CO_2 and a Carbon Tax. http://www.nocarbontaxrally.com/co2_the_truth.html.
5. The Big Wake Up. Climate Change Is Nothing but a Gigantic Fraud against the People! http://www.thebigwakeup.co.uk/Climate-Change/.
6. J. Tyndall. (1861). On the Absorption and Radiation of Heat by Gases and Vapours. *Philosophical Magazine*, Series 4, 22: 169–94, 273–85.
7. Skeptical Science. Rebuttal to "CO_2 Is Just a Trace Gas." http://www.skepticalscience.com/CO2-trace-gas.htm.
8. K. Braganza, D.J. Karoly, and J.M. Arblaster. (2004). Diurnal Temperature Range as an Index of Global Climate Change during the Twentieth Century. *Geophysical Research Letters*, 31: L13217.
9. Skeptical Science. Advanced Rebuttal to "It's Not Us." http://www.skepticalscience.com/its-not-us-advanced.htm.
10. M.J. Jarvis, B. Jenkins, and G.A. Rodgers. (1998). Southern Hemisphere Observations of a Long-Term Decrease in F Region Altitude and Thermospheric

Wind Providing Possible Evidence for Global Thermospheric Cooling. *Journal of Geophysical Research, Space Physics*, 103: 20775–787.

11. J. Laštovička et al. (2006). Global Change in the Upper Atmosphere. *Science*, 314: 1253–54.

12. Skeptical Science. Advanced Rebuttal to "CO_2 Effect Is Weak." http://www.skepticalscience.com/empirical-evidence-for-co2-enhanced-greenhouse-effect-advanced.htm.

13. Skeptical Science. Rebuttal to "CO_2 Increase Is Natural, Not Human-Caused." http://www.skepticalscience.com/co2-increase-is-natural-not-human-caused.htm.

14. Svante Arrhenius. (1896). On the Influence of Carbonic Acid in the Air upon the Temperature of the Ground. *Philosophical Magazine*, 41: 237–76.

15. Ian Sample. The Father of Climate Change. http://www.guardian.co.uk/environment/2005/jun/30/climatechange.climatechangeenvironment2.

16. Svante Arrhenius. (1908). *Worlds in the Making*. New York: Harper & Brothers.

17. R.R. Nemani et al. (2003). Climate-Driven Increases in Global Terrestrial Net Primary Production from 1982 to 1999. *Science*, 300: 1560–63.

18. A. Dai. (2011). Drought under Global Warming: A Review. *Wiley Interdisciplinary Reviews: Climate Change*, 2: 45–65. doi:10.1002/wcc.81.

19. Skeptical Science. Advanced Rebuttal to "CO_2 Is Plant Food." http://www.skepticalscience.com/co2-plant-food-advanced.htm.

20. Skeptical Science. Rebuttal to "It's Not Bad." http://www.skepticalscience.com/global-warming-positives-negatives-intermediate.htm.

21. Skeptical Science. OA Not OK Series. http://www.skepticalscience.com/search.php?Search=OAnotOKtextonlylogo.

22. InterAcademy Panel. (2009). IAP Statement on Ocean Acidification.

23. Skeptical Science. Rebuttal to "Ocean Acidification Isn't Serious." http://www.skepticalscience.com/ocean-acidification-global-warming.htm.

24. Skeptical Science. Rebuttal to "CO_2 Is Just a Trace Gas." http://www.skepticalscience.com/CO2-trace-gas.htm.

25. Spencer Weart and Raymond Pierrehumbert. A Saturated Gassy Argument. http://www.realclimate.org/index.php/archives/2007/06/a-saturated-gassy-argument/; http://www.realclimate.org/index.php/archives/2007/06/a saturated-gassy- argument-part-ii/.

26. John E. Harries et al. (2001). Increases in Greenhouse Forcing Inferred from the Outgoing Longwave Radiation Spectra of the Earth in 1970 and 1997. *Nature* 410, 355–57. doi:10.1038/35066553.

27. R. Philipona et al. (2004). Radiative Forcing—Measured at Earth's Surface—Corroborate the Increasing Greenhouse Effect. *Geophysical Research Letters*, 31, L03202: 1–4. doi:10.1029/2003GL018765.

28. W.F.J. Evans et al. (2006). *Measurements of the Radiative Surface Forcing of Climate*. 18th Conference on Climate Variability and Change, American Meteorological Society, January 30.

29. Skeptical Science. Rebuttal to "CO_2 Effect Is Saturated." http://www.skepticalscience.com/saturated-co2-effect.htm.

30. J. Inhofe. Floor Speeches. http://www.inhofe.senate.gov/newsroom/speech/climate-change-update.

31. View from Above. Interview with Marc Morano. http://viewhigh.blogspot.com/2007/08/is-earth-really-warming-part-2.html.

32. George Santayana. (1905). Reason in Common Sense. *The Life of Reason*, 1: 284.

CHAPTER 2

1. G.S. Callendar. (1938). The Artificial Production of Carbon Dioxide and Its Influence on Temperature. *Quarterly Journal Royal Meteorological Society*, 64: 223–40.

2. National Aeronautics and Space Administration (NASA) Goddard Institute for Space Studies (GISS). Measurements of Atmospheric CO_2 from Ice Cores and Direct Observations. http://data.giss.nasa.gov/modelforce/ghgases/Fig1A.ext.txt.

3. Callendar, The Artificial Production of Carbon Dioxide and Its Influence on Temperature.

4. Skeptical Science. Advanced Rebuttal to "It Warmed before 1940 When CO_2 Was Low." http://www.skepticalscience.com/global-warming-early-20th-century-advanced.htm.

5. J. Schlesinger. Climate Change: The Science Isn't Settled. http://www.ecolo.org/documents/documents_in_english/climate_change_Schlesinge.htm.

6. Gerald A. Meehl et al. (2004). Combinations of Natural and Anthropogenic Forcings in Twentieth-Century Climate. *American Meteorological Society Journal of Climate*, 17: 3721–27.

7. Gilbert N. Plass. (1956). The Carbon Dioxide Theory of Climatic Change. *Tellus*, 8: 140–54.

8. Gilbert N. Plass. (1956). Effect of Carbon Dioxide Variations on Climate. *American Journal of Physics*, 24: 376–87.

9. R. Knutti and G. Hegerl. (2008). The Equilibrium Sensitivity of the Earth's Temperature to Radiation Changes. *Nature Geoscience*, 1: 735–43.

10. Skeptical Science. Advanced Rebuttal to "Climate Sensitivity Is Low." http://www.skepticalscience.com/climate-sensitivity-advanced.htm.

11. A. Otto et al. (2013). Energy Budget Constraints on Climate Response. *Nature Geoscience*, 6: 415–16.

12. M.J. Ring et al. (2012). Causes of the Global Warming Observed since the 19th Century. *Atmospheric and Climate Sciences*, 2: 401.

13. N. Lewis. (2013). An Objective Bayesian Improved Approach for Applying Optimal Fingerprint Techniques to Estimate Climate Sensitivity. *Journal of Climate*, 26: 7414–29.

14. D.T. Shindell. (2014). Inhomogeneous Forcing and Transient Climate Sensitivity. *Nature Climate Change*, 4: 274–77.

15. J. Hansen, M. Sato, and R. Ruedy. (1997). Radiative Forcing and Climate Response. *Journal of Geophysical Research: Atmospheres (1984–2012)*, 102(D6): 6831–64.

16. J.R. Kummer and A.E. Dessler. (2014). The Impact of Forcing Efficacy on the Equilibrium Climate Sensitivity. *Geophysical Research Letters*, 41: 3565–68.

17. A.E. Dessler. (2010). A Determination of the Cloud Feedback from Climate Variations over the Past Decade. *Science*, 330(6010): 1523–27.

CHAPTER 3

1. National Center for Computational Sciences. Jaguar. http://www.nccs.gov/jaguar/.

2. S.I. Rasool and S.H. Schneider (1971). Atmospheric Carbon Dioxide and Aerosols: Effects of Large Increases on Global Climate. *Science*, 173, (3992): 138–41.

3. Skeptical Science. Intermediate Rebuttal to "Ice Age Predicted in the 70s." http://www.skepticalscience.com/ice-age-predictions-in-1970s-intermediate.htm.

4. T. Peterson et al. (2008). The Myth of the 1970s Global Cooling Scientific Consensus. 20th Conference on Climate Variability and Change, P4.7.

5. Pacific Northwest National Laboratory. (2004). Historical Sulfur Dioxide Emissions 1850–2000: Methods and Results. January.

6. Santayana, Reason in Common Sense, 284.

7. J.S. Sawyer. (1972). Man-Made Carbon Dioxide and the "Greenhouse" Effect. *Nature*, 239(5366): 2.

8. SMIC, I.C.M. (1971). Report of the Study of Man's Impact on Climate.

9. S. Manabe and R.T. Wetherald. (1967). Thermal Equilibrium of the Atmosphere with a Given Distribution of Relative Humidity. *Journal of Atmospheric Sciences*, 24: 241–59.

10. A.E. Dessler and S. Wong. (2009). Estimates of the Water Vapor Climate Feedback during El Niño-Southern Oscillation. *Journal of Climate*, 22(23): 6404–12.

11. Sawyer, Man-Made Carbon Dioxide and the "Greenhouse" Effect, 2.

12. W. Broecker. (1975). Climatic Change: Are We on the Brink of a Pronounced Global Warming? *Science*, 189(4201): 460–63.

13. M. Ahmed et al. (2013). Continental-Scale Temperature Variability during the Past Two Millennia. *Nature Geoscience*, 6: 339–46.

14. G. Feulner and S. Rahmstorf. (2010). On the Effect of a New Grand Minimum of Solar Activity on the Future Climate on Earth. *Geophysical Research Letters*, 37(5): L05707.

15. Gareth S. Jones, Mike Lockwood, and Peter A. Stott. (2012). What Influence Will Future Solar Activity Changes over the 21st Century Have on Projected Global Near-Surface Temperature changes? *Journal of Geophysical Research: Atmospheres (1984–2012)*, 117(D5).

16. J.G. Anet et al. (2013). Impact of a Potential 21st Century "Grand Solar Minimum" on Surface Temperatures and Stratospheric Ozone. *Geophysical Research Letters*, 40: 4420–25.

17. G.A. Meehl, J.M. Arblaster, and D.R. Marsh. (2013). Could a Future "Grand Solar Minimum" Like the Maunder Minimum Stop Global Warming? *Geophysical Research Letters*, 40: 1789–93.

18. NASA GISS. Global Temperature Trends: 2008 Annual Summation. http://data.giss.nasa.gov/gistemp/2008/.

19. A.P. Schurer, S.F. Tett, and G.C. Hegerl. (2014). Small Influence of Solar Variability on Climate over the Past Millennium. *Nature Geoscience*, 7(2): 104–8.

20. W. Kellogg. (1979). Influences of Mankind on Climate. *Annual Review of Earth and Planetary Sciences*, 7: 63.

21. Ibid.

22. Ibid.

23. J. Hansen et al. (1981). Climate Impact of Increasing Atmospheric Carbon Dioxide. *Science*, 213(4511): 957–66.

24. J. Hansen et al. (1988). Global Climate Changes as Forecast by Goddard Institute for Space Studies Three-Dimensional Model. *Journal of Geophysical Research*, 93: 9341–64. doi:10.1029/JD093iD08p09341.

25. A quote from Hansen's 1988 congressional testimony. http://www.cjr.org/essay/the_danger_of_fair_and_balance.php?page=all&print=true.

26. Brad Johnson. (2010). Cato's Pat Michaels Admits 40 Percent of Funding Comes from Big Oil. Climate Progress. http://thinkprogress.org/politics/2010/08/16/113717/oil-fueled-pat-michaels/.

27. P. Michaels. Kyoto Protocol: "A Useless Appendage to an Irrelevant Treaty." Testimony before the Committee on Small Business of the United States House of Representatives on July 29, 1998. http://www.cato.org/testimony/ct-pm072998.html.

28. Michael Crichton, *State of Fear*.

29. John Cook. The Quantum Theory of Climate Denial. http://skepticalscience.com/quantum-theory-of-climate-denial.html.

30. N. Oreskes. (2004). Beyond the Ivory Tower: The Scientific Consensus on Climate Change. *Science*, 306(5702): 1686.

31. P. Doran and M. Zimmerman. (2009). Examining the Scientific Consensus on Climate Change. *EOS, Transactions American Geophysical Union*, 90(3): 22.

32. W. Anderegg et al. (2010). Expert Credibility in Climate Change. *Proceedings of the National Academy of Sciences*, 107: 12107–9.

33. J. Cook et al. (2013). Quantifying the Consensus on Anthropogenic Global Warming in the Scientific Literature. *Environmental Research Letters*, 8(2): 024024.

34. Cato Institute. (2014). Richard Lindzen, Distinguished Senior Fellow, Center for the Study of Science. http://www.cato.org/people/richard-lindzen.

35. B. Ward. (2014). Blog on Select Committee Hearing. Grantham Research Institute on Climate Change and the Environment. http://www.lse.ac.uk/GranthamInstitute/news/blog-on-select-committee-hearing/.

36. W.D. Nordhaus. (2012). Why the Global Warming Skeptics Are Wrong. The New York Review of Books. http://www.nybooks.com/articles/archives/2012/mar/22/why-global-warming-skeptics-are-wrong/.

37. E. Mallove. (1989). Lindzen Critical of Global Warming Prediction. MIT Tech Talk. http://www.fortfreedom.org/s46.htm.

38. F. Guterl. The Truth about Global Warming. http://www.thedailybeast.com/newsweek/2001/07/22/the-truth-about-global-warming.html.

39. D. Nuccitelli, Lindzen Illusion #7: The Anti-Galileo. Skeptical Science. http://www.skepticalscience.com/lindzen-illusion-7-the-anti-galileo.html.

40. D. Nuccitelli. A Case Study of a Climate Scientist Skeptic. Skeptical Science. http://www.skepticalscience.com/a-case-study-of-a-climate-scientist-skeptic.html.

41. R.S. Lindzen and Y.S. Choi. (2009). On the Determination of Climate Feedbacks from ERBE Data. *Geophysical Research Letters*, 36(16): L16705.

42. K.E. Trenberth et al. (2010). Relationships between Tropical Sea Surface Temperature and Top-of-Atmosphere Radiation. *Geophysical Research Letters*, 37(3): L03702.

43. E.S. Chung, B.J. Soden, and B.J. Sohn. (2010). Revisiting the Determination of Climate Sensitivity from Relationships between Surface Temperature and Radiative Fluxes. *Geophysical Research Letters*, 37(10): L10703.

44. D.M. Murphy. (2010). Constraining Climate Sensitivity with Linear Fits to Outgoing Radiation. *Geophysical Research Letters*, 37(9): L09704.

45. A.E. Dessler. (2011). Cloud Variations and the Earth's Energy Budget. *Geophysical Research Letters*, 38(19): L19701.

46. C. Knappenberger. *Lindzen-Choi* "Special Treatment": Is Peer Review Biased against Nonalarmist Climate Science? MasterResource. http://www.masterresource.org/2011/06/lindzen-choi-special-treatment/.

47. R.S. Lindzen and Y.S. Choi (2011). On the Observational Determination of Climate Sensitivity and Its Implications. *Asia-Pacific Journal of Atmospheric Sciences*, 47(4): 377–90.

48. Richard Lindzen on WattsUpWithThat. http://wattsupwiththat.com/tag/richard-lindzen/.

49. E. Epstein. (2014). What Catastrophe? *The Weekly Standard*. http://www.weeklystandard.com/articles/what-catastrophe_773268.html.

50. CBS Boston. (2014). MIT Professor Urging Climate Change Activists to "Slow Down." http://boston.cbslocal.com/2014/01/14/mit-professor-urging-climate-change-activists-to-slow-down/.

51. N. Resnikoff. (2014). Bill Nye Spars with Member of Congress over Climate Change. MSNBC. Retrieved from http://www.msnbc.com/msnbc/bill-nye-spars-marsha-blackburn-meet-the-press.

CHAPTER 4

1. N. Oreskes. (2004). The Scientific Consensus on Climate Change. *Science*, 306(5702): 1686–1686.

2. T. Lambert. Peiser's 34 Abstracts. http://scienceblogs.com/deltoid/2005/05/06/peiser/.

3. Media Watch. Bolt's Minority View. http://www.abc.net.au/mediawatch/transcripts/s1777013.htm.

4. P.T. Doran and M.K. Zimmerman. (2009). Examining the Scientific Consensus on Climate Change. *Eos, Transactions American Geophysical Union*, 90(3): 22.

5. W.R. Anderegg et al. (2010). Expert Credibility in Climate Change. *Proceedings of the National Academy of Sciences*, 107(27): 12107–9.

6. Pew. (2012). More Say There Is Solid Evidence of Global Warming. Pew Research Center for the People and the Press. http://www.people-press.org/files/legacy-pdf/10-15-12%20Global%20Warming%20Release.pdf.

7. A. Leiserowitz et al. (2014). *Climate Change in the American Mind: April, 2014*. Yale University and George Mason University. New Haven, CT: Yale Project on Climate Change Communication.

8. http://research.greenpeaceusa.org/?a=view&d=2950.

9. F. Luntz. (2002). The Environment: A Cleaner, Safer, Healthier America. http://www.motherjones.com/files/LuntzResearch_environment.pdf.

10. D. Ding et al. (2011). Support for Climate Policy and Societal Action Are Linked to Perceptions about Scientific Agreement. *Nature Climate Change*, 1(9): 462–66.

11. S. Lewandowsky, G.E. Gignac, and S. Vaughan. (2012). The Pivotal Role of Perceived Scientific Consensus in Acceptance of Science. *Nature Climate Change*, 3(4): 399–404.

12. A. Leiserowitz et al. (2012). Climate Change in the American Mind: Americans' Global Warming Beliefs and Attitudes in March 2012. Yale University and George Mason University. New Haven, CT: Yale Project on Climate Change Communication. http://www.climatechangecommunication.org/images/files /Climate-Beliefs-March-2012.pdf.

13. M. Greenberg, D. Robbins, and S. Theel. (2013). STUDY: Media Sowed Doubt in Coverage of UN Climate Report. Media Matters for America. http: //mediamatters.org/research/2013/10/10/study-media-sowed-doubt-in coverage-of-un-clima/196387.

14. Union of Concerned Scientists. (2014). Science or Spin?: Assessing the Accuracy of Cable News Coverage of Climate Science. http://www.ucsusa.org /assets/documents/global_warming/Science-or-Spin-report.pdf.

15. The BBC Trust conclusions on the executive report on science impartiality review actions. http://www.bbc.co.uk/bbctrust/news/press_releases/2014 /science_impartiality.html.

16. K. Brysse et al. (2012). Climate Change Prediction: Erring on the Side of Least Drama? *Global Environmental Change*, 23: 327–37.

17. C. Heath and D. Heath. (2007). Made to Stick: Why Some Ideas Survive and Others Die. Random House Digital, Inc.

18. J. Cook and S. Lewandowsky. The Debunking Handbook. http://www .skepticalscience.com/docs/Debunking_Handbook.pdf.

19. A blog comment made by Richard Tol. https://andthentheresphysics. wordpress.com/2013/06/10/richard-tols-fourth-draft/.

20. D. Nuccitelli. Andrew Neil—These Are Your Climate Errors on BBC Sunday Politics. *The Guardian*. http://www.theguardian.com/environment /climate-consensus-97-per-cent/2013/jul/17/climate-change-scepticism andrew-neil-ed-davey.

21. Richard Tol has made the journal editor comments available on his blog here: http://richardtol.blogspot.com/2013/06/draft-comment-on-97-consensus paper.html#!/2013/06/draft-comment-on-97-consensus-paper.html.

22. Ibid.

23. D.R. Legates, W. Soon, and W.M. Briggs. (2013). Climate Consensus and "Misinformation": A Rejoinder to Agnotology, Scientific Consensus, and the Teaching and Learning of Climate Change. *Science & Education*, 1–20.

24. D. Bedford and J. Cook. (2013). Agnotology, Scientific Consensus, and the Teaching and Learning of Climate Change: A Response to Legates, Soon and Briggs. *Science & Education*, 22: 2019–30..

25. Energy Policy. (2014). Aims and Scope. http://www.journals.elsevier. com/energy-policy/.

26. Richard S. Tol. (2014). Quantifying the Consensus on Anthropogenic Global Warming in the Literature: A re-analysis. *Energy Policy*, 73: 701–5.

27. Ibid.

28. S.L. van der Linden et al. (2014). How to Communicate the Scientific Consensus on Climate Change: Plain Facts, Pie Charts or Metaphors? *Climatic Change*, 126: 255–62.

29. N. Stenhouse et al. (2013). Meteorologists' Views about Global Warming: A Survey of American Meteorological Society Professional Members. *Bulletin of the American Meteorological Society*, 95: 1029–40.

30. D. Nuccitelli. (2013). How Do Meteorologists Fit into the 97% Global Warming Consensus? The Guardian. http://www.theguardian.com/environment/climate-consensus-97-per-cent/2013/dec/02/meteorologists-global-warming-consensus.

CHAPTER 5

1. Intergovernmental Panel on Climate Change (IPCC) First Assessment Report, Working Group I Report. http://www.ipcc.ch/publications_and_data/publications_ipcc_first_assessment_1990_wg1.shtml.

2. National Oceanic and Atmospheric Administration (NOAA) Earth System Research Laboratory. Trends in Atmospheric Carbon Dioxide, Mauna Loa, Hawaii. http://www.esrl.noaa.gov/gmd/ccgg/trends/.

3. IPCC. Climate Change 1992: The Supplementary Report to the IPCC Scientific Assessment. http://www.ipcc.ch/publications_and_data/publications_ipcc_supplementary_report_1992_wg1.shtml.

4. IPCC Second Assessment Report, Working Group I Report. http://www.ipcc.ch/publications_and_data/publications_and_data_reports.shtml.

5. Don Easterbrook. Don Easterbrook's AGU Paper on Potential Global Cooling. http://wattsupwiththat.com/2008/12/29/don-easterbrooks-agu-paper on-potential-global-cooling/.

6. B. Hollingsworth. (2014). Climate Scientist Who Got It Right Predicts 20 More Years of Global Cooling. *CNS News*. http://cnsnews.com/news/article/barbara-hollingsworth/climate-scientist-who-got-it-right-predicts-20 more-years-global.

7. B. Angliss. (2010). Myth: Over 31,000 Scientists Signed the OISM Petition Project. Skeptical Science. http://www.skepticalscience.com/OISM-Petition-Project-intermediate.htm.

8. G. Musser. (2001). Climate of Uncertainty. *Scientific American*. http://www.scientificamerican.com/article/climate-of-uncertainty/.

9. IPCC Third Assessment Report, Working Group I Report. http://www.ipcc.ch/publications_and_data/publications_and_data_reports.shtml.

10. Skeptical Science. IEA CO_2 Emissions Update 2010—Bad News. http://www.skepticalscience.com/iea-co2-emissions-update-2010.html.

11. IPCC Fourth Assessment Report, Working Group I Report. http://www.ipcc.ch/publications_and_data/ar4/wg1/en/contents.html.

12. Skeptical Science. Why Wasn't the Hottest Decade Hotter? http://www.skepticalscience.com/Why-The-Hottest-Decade-Was-Not-Hotter-.html.

13. S.-I. Akasofu. Two Natural Components of the Recent Climate Change. http://people.iarc.uaf.edu/~sakasofu/pdf/two_natural_components_recent_climate_change.pdf.

14. S.-I. Akasofu. (2010). On the Recovery from the Little Ice Age. *Natural Science*, 2(11): 1211–24. http://www.scirp.org/Journal/PaperDownload.aspx?paperID=3217.

15. Skeptical Science. Advanced Rebuttal to "It's Cosmic Rays." http://www.skepticalscience.com/cosmic-rays-and-global-warming-advanced.htm.

16. W. Rea et al. (2011). Long Memory in Temperature Reconstructions. *Climatic Change*, 107(3–4): 247–65.

17. Skeptical Science. Advanced Rebuttal to "It's the Pacific Decadal Oscillation." http://www.skepticalscience.com/Pacific-Decadal-Oscillation-advanced.htm.

18. T. Crowley et al. (2000). Causes of Climate Change over the Past 1000 Years. *Science*, 289: 270.

19. W. Broecker. (2000). Was a Change in Thermohaline Circulation Responsible for the Little Ice Age? *Proceedings of the National Academy of Sciences*, 97(4): 1339–42.

20. W. Ruddiman. (2003). The Anthropogenic Greenhouse Era Began Thousands of Years Ago. *Climatic Change*, 61: 261–93.

21. Skeptical Science. Advanced Rebuttal to "We're Coming out of the LIA." http://www.skepticalscience.com/coming-out-of-little-ice-age-advanced.htm.

22. Syun-Ichi Akasofu. (2013). On the Present Halting of Global Warming. *Climate*, 1.1: 4–11.

23. C. Brierley. (2013). Why I Resigned from the Editorial Board of Climate over Its Akasofu Publication. Skeptical Science. http://skepticalscience.com/brierly-resignation-climate-akasofu.html.

24. Dana A. Nuccitelli et al. (2013). Comment on: Akasofu, S.-I. On the Present Halting of Global Warming. *Climate*, 1: 76–83.

25. C. Loehle and N. Scafetta. (2011). Climate Change Attribution Using Empirical Decomposition of Climatic Data. *Bentham Open Atmospheric Science Journal*, 5: 74–86.

26. R. Pierrehumber. How to Cook a Graph in Three Easy Lessons. http://www.realclimate.org/index.php/archives/2008/05/how-to-cook-a-graph in-three-easy-lessons/.

27. Attributed to von Neumann by Enrico Fermi, as quoted by Freeman Dyson. (January 22, 2004). A Meeting with Enrico Fermi. *Nature*, 427: 297.

28. C. Loehle and J.H. McCulloch. (2008). Correction to: A 2000-Year Global Temperature Reconstruction Based on Non-tree Ring Proxies. *Energy & Environment*, 19(1): 93–100.

29. A. Moberg et al. (2005). Highly Variable Northern Hemisphere Temperatures Reconstructed from Low- and High-Resolution Proxy Data. *Nature*, 433(7026): 613–17.

30. Q.B. Lu. (2013). Cosmic-Ray-Driven Reaction and Greenhouse Effect of Halogenated Molecules: Culprits for Atmospheric Ozone Depletion and Global Climate Change. *International Journal of Modern Physics B*, 27.

31. Waterloo News. (2013). Global Warming Caused by CFCs, Not Carbon Dioxide, Study Says. https://uwaterloo.ca/news/news/global-warming caused-cfcs-not-carbon-dioxide-study-says.

32. *Science Daily*. (2013). Global Warming Caused by CFCs, Not Carbon Dioxide, Researcher Claims in Controversial Study. http://www.sciencedaily.com/releases/2013/05/130530132443.htm.

33. D. Nuccitelli et al. (2014). Comment on "Cosmic-Ray-Driven Reaction and Greenhouse Effect of Halogenated Molecules: Culprits for Atmospheric Ozone Depletion and Global Climate Change." *International Journal of Modern Physics B*, 28(13).

34. J. McLean. Statement: COOL YEAR PREDICTED: Updated with LATEST GRAPH. http://climaterealists.com/index.php?id=7349.

35. J. McLean et al. (2009). Influence of the Southern Oscillation on Tropospheric Temperature. *Journal of Geophysical Research*, 114, D14104: 8.

36. J. McLean. Comments on Carbon Tax Is Politically Astute but Profoundly Inadequate. http://theconversation.edu.au/carbon-tax-plan-is-politically-astute-but-profoundly-inadequate-1975#comment_5207.

37. J. McLean. (2014). How Politics Clouds the Climate Change Debate. *The Age*. http://www.theage.com.au/comment/how-politics-clouds-the-climate change-debate-20140102-307ja.html.

CHAPTER 6

1. Skeptical Science. The Escalator. http://www.skepticalscience.com/graphics.php?g=47.

2. K. Cowtan and R.G. Way. (2013). Coverage Bias in the HadCRUT4 Temperature Series and Its Impact on Recent Temperature Trends. *Quarterly Journal of the Royal Meteorological Society*, 140: 1935–44.

3. Y. Kosaka and S.P. Xie. (2013). Recent Global-Warming Hiatus Tied to Equatorial Pacific Surface Cooling. *Nature*, 501: 403–7.

4. J.S. Risbey et al. (2014). Well-Estimated Global Surface Warming in Climate Projections Selected for ENSO Phase. *Nature Climate Change*, 4: 835–40.

5. M. Huber and R. Knutti. (2014). Natural Variability, Radiative Forcing and Climate Response in the Recent Hiatus Reconciled. *Nature Geoscience*, 7, 651–56.

6. M. Watanabe et al. (2013). Strengthening of Ocean Heat Uptake Efficiency Associated with the Recent Climate Hiatus. *Geophysical Research Letters*, 40: 3175–79.

7. G.A. Meehl et al. (2011). Model-Based Evidence of Deep-Ocean Heat Uptake during Surface-Temperature Hiatus Periods. *Nature Climate Change*, 1: 360–64.

8. Gerald A. Meehl et al. (2013). Externally Forced and Internally Generated Decadal Climate Variability Associated with the Interdecadal Pacific Oscillation. *Journal of Climate*, 26: 7298–310.

9. V. Guemas et al. (2013). Retrospective Prediction of the Global Warming Slowdown in the Past Decade. *Nature Climate Change*, 3: 649–53.

10. S. Levitus et al. (2012). World Ocean Heat Content and Thermosteric Sea Level Change (0–2000 m), 1955–2010. *Geophysical Research Letters*, 39: L10603.

11. D.H. Douglass and R.S. Knox. (2012). Ocean Heat Content and Earth's Radiation Imbalance. II. Relation to Climate Shifts. *Physics Letters A*, 376(14), 1226–29.

12. D. Nuccitelli et al. (2012). Comment on "Ocean Heat Content and Earth's Radiation Imbalance. II. Relation to Climate Shifts." *Physics Letters A*, 376: 3466–68

13. M.A. Balmaseda, K.E. Trenberth, and E. Källén. (2013). Distinctive Climate Signals in Reanalysis of Global Ocean Heat Content. *Geophysical Research Letters*, 40: 1754–59.

14. M.H. England et al. (2014). Recent Intensification of Wind-Driven Circulation in the Pacific and the Ongoing Warming Hiatus. *Nature Climate Change*, 4: 222–27.

15. M. Lott and C. Couger. Draft UN Climate Report Shows 20 Years of Overestimated Global Warming, Skeptics Warn. *Fox News* (January 28, 2013). http://www.foxnews.com/science/2013/01/28/un-climate-report-models overestimated-global-warming/.

16. Tamino. Open Mind. (December 20, 2012). Fake Skeptic Draws Fake Picture of Global Temperature. http://tamino.wordpress.com/2012/12/20/fake-skeptic-draws-fake-picture-of-global-temperature/.

17. D. Nuccitelli. (October 1, 2013). IPCC Model Global Warming Projections Have Done Much Better Than You Think. *The Guardian*. http://www.theguardian.com/environment/climate-consensus-97-per-cent/2013/oct/01/ipcc-global-warming-projections-accurate.

18. S. McIntyre. Climate Audit. (September 30, 2013). IPCC: Fixing the Facts. http://climateaudit.org/2013/09/30/ipcc-disappears-the-discrepancy/.

19. M.E. Mann, R.S. Bradley, and M.K. Hughes. (1998). Global-Scale Temperature Patterns and Climate Forcing over the Past Six Centuries. *Nature*, 392(6678): 779–87.

20. M.E. Mann, R.S. Bradley, and M.K. Hughes. (1999). Northern Hemisphere Temperatures during the Past Millennium: Inferences, Uncertainties, and Limitations. *Geophysical Research Letters*, 26(6): 759–62.

21. S. McIntyre and R. McKitrick. (2005). Hockey Sticks, Principal Components, and Spurious Significance. *Geophysical Research Letters*, 32: L03710.

22. M. Ahmed et al. (2013). Continental-Scale Temperature Variability during the Past Two Millennia. *Nature Geoscience*, 6: 339–46.

23. M.E. Mann. (2012). *The Hockey Stick and the Climate Wars*. New York: Columbia University Press.

24. R. McKitrick. Climate Audit (2013). http://climateaudit.org/2013/09/30/ipcc-disappears-the-discrepancy/#comment-441605.

25. S. Rahmstorf et al. (2007). Recent Climate Observations Compared to Projections. *Science*, 316(5825), 709–709.

26. Media Matters for America. (2013). STUDY: Media Sowed Doubt in Coverage of UN Climate Report. http://mediamatters.org/research/2013/10/10/study-media-sowed-doubt-in-coverage-of-un-clima/196387.

27. The IPCC Fifth Assessment Report (AR5), Working Group I Summary for Policymakers (SPM). http://www.ipcc.ch/report/ar5/wg1/.

28. E. Massod. (2013). Climate Change Sceptics Aren't All Alike, So Don't Tar Them with the Same Brush. http://www.theguardian.com/environment/2013/oct/09/climate-change-sceptics-not-all-alike?CMP=twt_gu.

29. M. Hoerling et al. (2012). On the Increased Frequency of Mediterranean Drought. *Journal of Climate*, 25: 2146–61.

30. The IPCC Fifth Assessment Report (AR5), Working Group II Summary for Policymakers (SPM). http://www.ipcc.ch/report/ar5/wg2/.

31. Ibid.

32. F. Ackerman and C. Munitz. (2012). Climate Damages in the FUND Model: A Disaggregated Analysis. *Ecological Economics*, 77: 219–24.

33. D.I. Stern. (2012). Letter from the Associate Editor Concerning the Comments from Anthoff and Tol and Ackerman and Munitz. *Ecological Economics*, 81: 41–41.

34. J. Kuylenstierna and J. Rockström. (2013). Statement from SEI Leadership. http://frankackerman.com/Tol/SEI_Statement.pdf.

35. United Nations Environment Programme. (2014). A Changing Climate Creates Pervasive Risks but Opportunities Exist for Effective Responses—IPCC Report. http://www.unep.org/Documents.Multilingual/Default.asp?DocumentID=2764&ArticleID=10773&l=en.

36. The IPCC Fifth Assessment Report (AR5), Working Group III Summary for Policymakers (SPM). http://www.ipcc.ch/report/ar5/wg3/.

37. G. Lloyd. (2014). Carbon Cure Cost More Painful Than the Illness, Says Bjorn Lomborg. http://www.theaustralian.com.au/national-affairs/policy/carbon-cure-cost-more-painful-than-the-illness-says-bjorn-lomborg/story-e6frg6xf-1226884300046#.

38. D. Nuccitelli. (2014). Climate Dollars and Sense—Preventing Global Warming Is the Cheap Option. *The Guardian.* http://www.theguardian.com/environment/climate-consensus-97-per-cent/2014/apr/22/preventing-global-warming cheaper-than-adapting.

39. W. Nordhaus and P. Sztorc. (2013). DICE 2013: Introduction and User's Manual. http://www.econ.yale.edu/~nordhaus/homepage/documents/Dicemanualfull.pdf.

40. P. Krugman. (2014). Salvation Gets Cheap. http://www.nytimes.com/2014/04/18/opinion/krugman-salvation-gets-cheap.html.

41. Santayana, Reason in Common Sense, 284.

42. J. Mashey. (2011). Skeptics Prefer Pal Review over Peer Review: Chris de Freitas, Pat Michaels and Their Pals, 1997–2003. DeSmogBlog. http://www.desmogblog.com/skeptics-prefer-pal-review-over-peer-review-chris-de-freitas-pat-michaels-and-their-pals-1997-2003.

43. W. Soon and S. Baliunas. (2003). Proxy Climatic and Environmental Changes of the Past 1000 Years. *Climate Research*, 23(2): 89–110.

44. M. Mann et al. (2003). On Past Temperatures and Anomalous Late-20th Century Warmth. *Eos, Transactions American Geophysical Union*, 84(27): 256–256.

45. J. Inhofe. Floor Speeches. http://www.inhofe.senate.gov/newsroom/speech/climate-change-update.

46. Logical Science. (n.d.). Sallie Baliunas & Willie Soon. www.odlt.org/dcd/docs/logical%20science_Sallie%20Baliunas.pdf.

47. R.W. Spencer and W.D. Braswell. (2011). On the Misdiagnosis of Surface Temperature Feedbacks from Variations in Earth's Radiant Energy Balance. *Remote Sensing*, 3(8): 1603–13.

48. K. Trenberth and J. Fasullo. (2011). Misdiagnosis of Surface Temperature Feedback.RealClimate.http://www.realclimate.org/index.php/archives/2011/07/.

49. W. Wagner. (2011). Taking Responsibility on Publishing the Controversial Paper "On the Misdiagnosis of Surface Temperature Feedbacks from Variations in Earth's Radiant Energy Balance" by Spencer and Braswell. *Remote Sensing*, 3(8): 1603–13; *Remote Sensing*, 3(9): 2002–4.

50. K. Trenberth. (2011). Roy Spencer's Paper on Climate Sensitivity. Skeptical Science. http://www.skepticalscience.com/roy-spencer-negative-feedback-climate-sensitivity-advanced.htm.

51. K.E. Trenberth, J.T. Fasullo, and J.P. Abraham. (2011). Issues in Establishing Climate Sensitivity in Recent Studies. *Remote Sensing*, 3(9): 2051–56.

52. R.S. Lindzen and Y.S. Choi. (2009). On the Determination of Climate Feedbacks from ERBE Data. *Geophysical Research Letters*, 36(16): L16705.

53. K.E. Trenberth et al. (2010). Relationships between Tropical Sea Surface Temperature and Top-of-Atmosphere Radiation. *Geophysical Research Letters*, 37(3): L03702.

54. J. Gillis. (2012). Clouds' Effect on Climate Change Is Last Bastion for Dissenters. *The New York Times*. http://www.nytimes.com/2012/05/01/science/earth/clouds-effect-on-climate-change-is-last-bastion-for-dissenters.html.

55. R.W. Spencer and W.D. Braswell. (2013). The Role of ENSO in Global Ocean Temperature Changes during 1955–2011 Simulated with a 1D Climate Model. *Asia-Pacific Journal of Atmospheric Sciences*, 50: 229–37.

56. J. Fong. (2011). Fox Tries to Debunk Global Warming, Fails Miserably. Media Matters for America. http://mediamatters.org/research/2011/01/27/fox-tries-to-debunk-global-warming-fails-misera/183174.

57. R.W. Spencer. (2011). FUNDANOMICS: The Free Market, Simplified. http://www.drroyspencer.com/2011/07/fundanomics-the-free-market-simplified/.

58. R.W. Spencer. (2014). Time to Push Back against the Global Warming Nazis. http://www.drroyspencer.com/2014/02/time-to-push-back-against-the global-warming-nazis/.

59. G. Monbiot. (2011). The Spectator Runs False Sea-Level Claims on Its Cover. *The Guardian*. http://www.theguardian.com/environment/georgemonbiot/2011/dec/02/spectator-sea-level-claims.

60. Scholarly Open Access. (2013). Recognizing a Pattern of Problems in "Pattern Recognition in Physics." http://scholarlyoa.com/2013/07/16/recognizing-a-pattern-of-problems-in-pattern-recognition-in-physics/.

61. M. Rasmussen. (2014). Termination of the journal *Pattern Recognition in Physics*. http://www.pattern-recognition-in-physics.net/.

62. M. Rasmussen. (2014). Termination of the Journal *Pattern Recognition in Physics*. http://www.pattern-recognition-in-physics.net/.

63. P. Thornton. (2013). The Obamacare Exemptions That Aren't. *LA Times*. http://www.latimes.com/opinion/letters/la-le-1005-shutdown-obamacare mailbag-20131005,0,3748083.story#axzz2r5xtNQSW.

64. P. Thornton. (2013). On Letters from Climate-Change Deniers. *LA Times*. http://www.latimes.com/opinion/opinion-la/la-ol-climate-change-letters 20131008,0,871615.story#axzz2r5xtNQSW.

65. J. Lewis and M. McEvoy. (2013). Climate Change: A Note from Our Letters Editors. *The Sydney Morning Herald*. http://www.smh.com.au/comment/smh-letters/climate-change-a-note-from-our-letters-editors-20131021–2vvjd.html.

66. G. Readfearn. (2013). Climate Sceptic Group Reveals Strategy Document to Win Hearts and Minds. *DeSmogBlog*. http://www.desmogblog.com/climate-sceptic-group-reveals-strategy-document-win-hearts-and-minds.

CHAPTER 7

1. A. Barnosky. (2011). Has the Earth's Sixth Mass Extinction Already Arrived? *Nature*, 471: 51–57.

2. Skeptical Science. The Earth's Sixth Mass Extinction May Be Underway. http://www.skepticalscience.com/sixth-mass-extinction.html.

3. Skeptical Science. Monckton Myth #5: Dangerous Warming. http://www.skepticalscience.com/monckton-myth-5-dangerous-warming.html.

4. Australian Climate Commission. The Critical Decade 2013. https://climatecommission.wordpress.com/.

5. S. Pacala and R. Socolow. (2004). Stabilization Wedges: Solving the Climate Problem for the Next 50 Years with Current Technologies. *Science*, 305(5686): 968–72.

6. J. Romm. Is 450 ppm (or Less) Politically Possible? Part 1. http://thinkprogress.org/romm/2008/03/31/202489/is-450-ppm-carbon-dioxide politically-possible-1/.

7. R. Socolow. Wedges Reaffirmed. http://www.thebulletin.org/web edition/features/wedges-reaffirmed.

8. J. Hansen. The High Cost of Inaction. http://www.realclimate.org/index.php/archives/2011/10/the-cost-of-inaction/.

9. N. Chestney. Warming Limit Risk If No Climate Action by 2017: IEA. http://www.reuters.com/article/2011/11/11/us-climate-iea-idUSTRE7AA38C20111111.

10. P. Krugman. (2011). Markets Can Be Very, Very Wrong. *New York Times*. http://krugman.blogs.nytimes.com/2011/09/30/markets-can-be-very-very-wrong/.

11. A. Benjamin. (2007). Stern: Climate Change a "Market Failure." *The Guardian*. http://www.theguardian.com/environment/2007/nov/29/climatechange.carbonemissions.

12. R.S. Tol. (2009). The Economic Effects of Climate Change. *The Journal of Economic Perspectives*, 23(2): 29–51.

13. W.D. Nordhaus. (2012). Why the Global Warming Skeptics Are Wrong. *The New York Review of Books*. http://www.nybooks.com/articles/archives/2012/mar/22/why-global-warming-skeptics-are-wrong/.

14. C. Hope. (2011). The Social Cost of CO_2 from the PAGE09 Model. *Economics Discussion Paper*, 2011–39.

15. F. Ackerman and E.A. Stanton. (2010). The Social Cost of Carbon. Economics for Equity and Environment (E3 Network). http://www. e3network. org/papers/SocialCostOfCarbon_SEI_20100401.pdf.

16. J. Samson et al. (2011). Geographic Disparities and Moral Hazards in the Predicted Impacts of Climate Change on Human Populations. *Global Ecology and Biogeography*, 20(4): 532–44.

17. R. Tol. (2014). The Economic Impact of Climate Change in the 20th and 21st Centuries. Copenhagen Consensus Center. http://www.copenhagenconsensus.com/sites/default/files/climate_change.pdf.

18. B. Plumer. (2013). An Obscure New Rule on Microwaves Can Tell Us a Lot about Obama's Climate Policies. *The Washington Post* Wonkblog. http://www.washingtonpost.com/blogs/wonkblog/wp/2013/06/05/what-an-obscure-microwave-rule-says-about-obamas-climate-plans/.

19. National Geographic. The Great Energy Challenge. http://environment.nationalgeographic.com/environment/energy/great-energy-challenge/global-energy-subsidies-map/.

20. C. Keating. 20 Most Profitable Companies. CNN, Fortune & Money. http://money.cnn.com/galleries/2012/fortune/1205/gallery.500-most-profitable. fortune/.

21. R. Conniff. (2009). The Political History of Cap-and-Trade. *Smithsonian* Magazine. http://www.smithsonianmag.com/science-nature/Presence-of-Mind-Blue-Sky-Thinking.html.

22. U.S. Environmental Protection Agency. (2011). The Benefits and Costs of the Clean Air Act from 1990 to 2020. http://www.epa.gov/air/sect812/feb11/summaryreport.pdf.

23. D.S. Elgie and J.A. McClay. (2013). BC's Carbon Tax Shift after Five Years: Results. *Sustainable Prosperity*. http://www.sustainableprosperity.ca/dl1026&display.

24. The Environics Institute. (2012). Focus Canada 2012—Climate Change: Do Canadians Still Care? http://www.environicsinstitute.org/institute-projects/completed-projects/focus-canada-2012.

25. Regional Economic Models, Inc. (REMI) and Synapse Energy Economics, Inc. (2014). The Economic, Climate, Fiscal, Power, and Demographic Impact of a National Fee-and-Dividend Carbon Tax. Retrieved from http://citizensclimatelobby.org/wp-content/uploads/2014/06/REMI-carbon-tax-report-62141.pdf.

26. Clean Air Act, US Code of Federal Regulations, Title 42, Section 7408 http://www.law.cornell.edu/uscode/text/42/7408.

27. US EPA. (2003). EPA Denies Petition to Regulate Greenhouse Gas Emissions from Motor Vehicles. http://yosemite.epa.gov/opa/admpress.nsf/fb36d84bf0a1390c8525701c005e4918/694c8f3b7c16ff6085256d900065fdad!OpenDocument.

28. US EPA (2009). Endangerment and Cause or Contribute Findings for Greenhouse Gases under Section 202(a) of the Clean Air Act. http://www.epa.gov/climatechange/endangerment/.

29. FindLaw. Coalition for Responsible Regulation, Inc., et al., Petitioners v. Environmental Protection Agency, Respondent. http://caselaw.findlaw.com/us-dc-circuit/1604469.html.

30. US EPA. (2013). Vehicle Standards and Regulations. http://www.epa.gov/otaq/standards.htm.

31. US EPA. (2013). 2013 Proposed Carbon Pollution Standard for New Power Plants. http://www2.epa.gov/carbon-pollution-standards/2013-proposed-carbon-pollution-standard-new-power-plants.

32. S. Cardoni. (2013). Obama to Congress: Act on Climate Change, or I Will. *Yahoo News*. http://news.yahoo.com/obama-congress-act-climate-change-155810445.html.

33. S. Jones. (2013). Kerry: Obama Will Act on Climate Change without Congressional Approval. *CSN News*. http://cnsnews.com/news/article/kerry-obama-will-act-climate-change-without-congressional-approval.

34. Energy and Enterprise Initiative. (2014). Putting Free Enterprise to Work on Energy and Climate. http://energyandenterprise.com/.

35. The White House. (2013). The President's Climate Action Plan. http://www.whitehouse.gov/sites/default/files/image/president27sclimateactionplan.pdf.

36. A. Freedman. (2014). Obama's State of the Union Climate Mention Fits Pattern. *Climate Central*. http://www.climatecentral.org/news/obamas-latest-state-of-the-union-climate-mention-fits-pattern-17007.

37. California Environmental Protection Agency Air Resources Board. Cap-and-Trade. http://arb.ca.gov/cc/capandtrade/capandtrade.htm.

38. *Word News Australia*. Factbox: Carbon Taxes around the World. http://www.sbs.com.au/news/article/1492651/factbox-carbon-taxes-around-the-world.

39. C. Johnson. (2013). Most Want to Keep Carbon Tax: Poll. *The Sydney Morning Herald*. http://www.smh.com.au/federal-politics/political-news/most-want-to-keep-carbon-tax-poll-20130713-2px4d.html.

40. M. Griffin. (2013). Anti-Carbon Price Party Wins Australian Election; New Senate Likely to Favor Repeal. *Bloomberg BNA*. http://www.bna.com/anticarbon-price-party-n17179876974/.

41. Gerard Wynn. (2014). China Completes Rollout of Pilot Carbon Markets. Responding to Climate Change. http://www.rtcc.org/2014/06/19/china-completes-rollout-of-pilot-carbon-markets/.

42. L.T. Johnson and C. Hope. (2012). The Social Cost of Carbon in US Regulatory I.mpact Analyses: An Introduction and Critique. *Journal of Environmental Studies and Sciences*, 2(3): 205–21.

43. Union of Concerned Scientists. State of Charge: Electric Vehicles' Global Warming Emissions and Fuel-Cost Savings across the United States. http://www.ucsusa.org/clean_vehicles/smart-transportation-solutions/advanced-vehicle-technologies/electric-cars/emissions-and-charging-costs-electric-cars.html.

44. D. Undercoffler. (2013). Tesla Motors Plans to Debut Cheaper Car in Early 2015. *Los Angeles Times*. http://articles.latimes.com/2013/dec/15/autos/la-fi-hy-autos-tesla-model-e-debut-2015-20131213.

45. Think Progress. US Responsible for 29% of Carbon Dioxide Emissions over Past 150 Years, Triple China's Share. http://thinkprogress.org/romm/2009/06/01/204179/us-responsible-for-29-of-carbon-dioxide-emissions-over-past-150-years-triple-chinas-share/.

46. George Mason University Center for Climate Change Communication. (2011). Politics and Global Warming; Democrats, Republicans, Independents, and the Tea Party. http://www.climatechangecommunication.org/images/files/PoliticsGlobalWarming2011.pdf.

47. L. Feldman et al. (2014). The Mutual Reinforcement of Media Selectivity and Effects: Testing the Reinforcing Spirals Framework in the Context of Global Warming. *Journal of Communication*, 64(4): 590–611.

48. C. Sagan. (1980). *Broca's Brain: Reflections on the Romance of Science*. Random House LLC.

49. F. Guterl. (2001). The Truth about Global Warming. *Newsweek*. http://www.newsweek.com/truth-about-global-warming-154937.

50. J. Romm. (2014). Quoting John Christy on Climate Change Is Like Quoting Dick Cheney on Iraq. Climate Progress. Retrieved from http://thinkprogress.org/climate/2014/07/17/3461320/john-christy-climate-change-dick-cheney-iraq/.

51. Interview on Real Time with Bill Maher on February 4, 2011. http://www.hbo.com/real-time-with-bill-maher/episodes/0/201-episode/synopsis/quotes.html#/.

Index

Note: Page numbers followed by *"f"* indicate a figure on that page.

About the Author

DANA NUCCITELLI is an environmental scientist and risk assessor at a private environmental consulting firm in California. He is a climate blogger for *The Guardian* and *Skeptical Science* and has published several peer-reviewed climate papers. Nuccitelli holds a bachelor's degree in astrophysics from the University of California at Berkeley and a master's degree in physics from the University of California at Davis.